Signals, Oscillations, and Waves

A Modern Approach

For a complete listing of the *Artech House Signal Processing Library*,
turn to the back of this book.

Signals, Oscillations, and Waves

A Modern Approach

David Vakman

Artech House
Boston • London

Library of Congress Cataloging-in-Publication Data
Vakman, D. E.
Signals, oscillations, and waves: a modern approach / David Vakman.
p. cm. — (Artech House signal processing library)
Includes bibliographical references and index.
ISBN 0-89006-814-3 (alk. paper)
1. Signal processing—Mathematics.
2. Radio waves—Mathematical models. I. Title. II. Series.
TK5102.9.V34 1998
621.382'2—dc21 98-6460
CIP

British Library Cataloguing in Publication Data
Vakman, David
Signals, oscillations, and waves: a modern approach— (Artech House signal processing library)
1. Signal processing 2. Waves 3. Oscillations
I. Title
621.3'822

ISBN 0-89006-814-3

Cover design by Jennifer Stuart

685 Canton Street
Norwood, MA 02062

International Standard Book Number: 0-89006-814-3
Library of Congress Catalog Card Number: 98-6460

10 9 8 7 6 5 4 3 2 1

In memory of Lev Albertovich Wainstein,
whose ideas penetrate this book

Contents

Preface

What is this book about? In short, it studies amplitudes and frequencies of signals and waves and shows how they work in engineering and physical problems.

The notions of amplitude and frequency in their simplest form are known and widely employed in engineering. Their simplicity, however, is illusory.

Operating with the strict notions of amplitude and frequency is not an easy thing to do. A rigorous approach employed in this book may be conflicting in certain points, but it does work. Going deeply into the amplitude and frequency notions, we solve many important problems. Even classical problems investigated long ago will be solved more efficiently.

If you have a taste for theory and are interested in such fields as signal processing, electrical or electronic engineering, mechanical or seismic oscillations, wave propagation and many others, you will find in this book a tool to be used in your research. At least, it often works better than some of the methods currently applied. So go ahead.

1

Introduction

Light, sound, radio signals, quantum particles, and many others are *oscillation phenomena* studied by physicists and engineers since the 19th century. The advanced theory of oscillations and waves has been developed, and its basic methods are well known. So why reconsider the known problems and explore the fundamental characteristics of signals and waves—their amplitude and frequency?

A similar situation occurred in optics. For a long time, light was interpreted as rays, and optical phenomena, such as reflection and refraction, was realized within ray representations. It was then discovered that light is also a wave, and the other phenomena, such as diffraction or dispersion, are wavelike in nature. Local rays and global waves are very diverse, and it is amazing that both represent the same physical object. Nevertheless, it was also shown that the global waves are transformed into local rays if a wavelength is very short. Therefore, the global lightwaves do not conflict with but generalize the local rays. In fact, light is a phenomenon of dual nature—local and global.

A main point of this book is that amplitude and frequency of signals and waves are also of dual nature and can be represented by local and global concepts. Commonly, we believe that the signal amplitude can be modulated and detected at a given instant. This local approach is fundamental for radio communications: nobody knows beforehand what the announcer will say at a given instant, and therefore, modulation and detection must be local procedures. In the strict sense, however, amplitude and frequency are global, and we have to know the whole signal from $t = -\infty$ to $t = \infty$ for defining them for a given instant. We intend to show that:

- The global approach to amplitude and frequency generalizes common local representations.

- Using this approach, we deeply understand oscillation and wave phenomena. We also obtain more accurate results or solve new problems unattainable for local methods.

We will show this for various problems related to many fields, and we often prefer old problems for cogency.

A list of the main problems considered in this book follows. The book [1] written 15 years ago by L. A. Wainstein and the author contains some of these problems. Since then, however, most of them have been revised, and many results are presented here for the first time.

Analytic Signals Global amplitude and frequency are related to the *analytic signal* (AS) introduced by D. Gabor in 1946. For a given real signal $u(t) = a(t) \times \cos\phi(t)$, the amplitude and phase are ambiguous because we have only one equation for two unknowns—$a(t)$ and $\phi(t)$. Therefore, even if a real signal is completely known, its amplitude and phase are inexplicit. The AS is a *complex* signal $w(t) = a(t)e^{i\phi(t)}$ that defines the amplitude and phase unambiguously. The AS, however, is a global method, and the whole signal $u(t)$ must be known at $-\infty < t < \infty$ for defining the $w(t)$ at a given t. Therefore, it is unclear how amplitudes or frequencies given by the AS are detected and modulated within a short time.

Local and Global Representations Interrelations between local and global approaches are discussed in detail. Two kinds of generators are employed for chirp signals. One of them is local in the sense that a signal is formed in a short time around each instant, whereas another generator is global. In spite of this distinction, the signals produced with both generators are close or identical. We consider a condition for their equivalence. We also consider paradoxes of instantaneous frequency and early attempts of frequency modulation. The attempts were unsuccessful because the condition was violated.

Alternative Amplitudes and Frequencies Since the 1930s, alternative local amplitudes and frequencies discrepant from the AS have been suggested by some authors. Today, ample potentialities of signal processing also stimulate the search of new amplitudes and phases adjusted to modern methods. We will show, however, that only the AS meets reasonable physical conditions. All alternative methods violate the conditions and result in improper answers.

Radio-Engineering Devices Practical radio devices—modulators, detectors, mixers, and the like—were invented long before the AS. Nevertheless, the amplitudes and frequencies employed agree with the AS. A basic technical idea was *spectral separation* of modulation from a carrier with filters. Therefore, their spectra are nonoverlapping, and the amplitude and frequency extracted or transformed in radio devices are the same as given by the AS. So, in spite of its paradox, the AS is widely used and reasserted with radio-engineering practice.

Noise Frequency Modulation and Diffusion In theory of random signals, the AS is widely accepted. It provides elegant relations between correlation functions, but the local frequency idea is also applied to random signals.

According to the local frequency idea, we expect that the power spectrum of frequency modulation follows the probability of a modulating noise. This is true for slow modulation, whereas fast modulation results in the global spectrum of diffusion. So, random modulation and diffusion are associated with the local and global frequency notions.

Frequency Stability Measurements This problem is the most practical in the book. Frequency stability is very important for engineering, physics, and astronomy. Up to now, however, stability measurements are based on the local approach, and *short-term instability* is necessarily measured during a short time. Even for highly stable low-noise generators, this results in significant measuring errors. For global measurements based on the AS, short-term instability within 1 ms is measured during a *long* time of about 1 sec, but frequency variations are obtainable for 1 ms. The advantage of accuracy over local methods is great, and noise errors are practically eliminated.

Generator Theory Studying a generator in the 1920s, B. Van der Pol originated the averaging method. Since then, averaging is often used for defining amplitude and frequency of nonlinear oscillations. Removing higher harmonics, however, averaging ignores an important cause of instability and is true for the first order only.

The AS provides a more general method for nonlinear and parametric oscillations that improves many known results. We will find out that *dynamic* frequency fluctuations accompany amplitude variations in a generator. This second-order effect explains flicker frequency instability. More accurate results are also obtained for thermal and shot noise fluctuations.

Using *complex equations* for the AS obtained from initial real equations, we also develop the *frequency separation procedure*. This procedure allows us to study nonlinear interaction of harmonics and find higher-order effects in multiharmonic oscillation systems.

Other Oscillation Systems We also find new solutions for many classical oscillation systems. So, for a pendulum of varying length, the *momentum* is conserved in higher orders instead of the *adiabatic invariant* known for the first order. Also, nonlinear and parametric oscillators show very different behavior for low-frequency, resonant, and high-frequency excitation. The AS and frequency separation provide powerful methods for such systems.

Electron Motion In electric and magnetic fields, the electron runs on a circular orbit while the orbit slowly drifts along the equipotential line. This motion is fundamental for magnetrons and other electronic devices, and in the first order, it has been studied with averaging. Using the frequency separation procedure, we explore the motion in detail and find out many additional effects. So, the circular orbit becomes elliptic, and its size is varying in a nonuniform electric field. Velocity of rotation is also varying, and energy of the orbital motion is conserved. Besides, small epicycles appear with fast reverse rotations, and the drift deviates from the equipotential.

Wave Phenomena For wave phenomena, *analytic waves* (AWs) play the same role as analytic signals. Using the AW, we define amplitudes and frequencies of real running waves. Also, the wave center and duration are the same for real waves and their AW. The main result obtained is that local group delay averaged in frequency defines velocity of a wave center at each point. In uniform media, the center moves with a constant velocity, whatever dispersion.

A paradoxical phenomenon occurs in damping media where velocity of the wave center may exceed light velocity. This paradox, however, has a simple explanation for frequency modulated waves, and no physical conflict arises.

Using the AW, we modify Whitham's method for defining not only frequency but also amplitude of a wave. The wave shape depends on interaction between group delays in the initial spectrum and in the medium. They may compensate each other, and the wave becomes compressed in time. We also develop an asymptotic method of higher accuracy than Whitham's method.

Quantum Mechanical Wave Packets Schrödinger's equation defines wave packets moving in nonuniform dispersive media, and the wave center represents the associated classical particle. A wave packet is often the AW. This comes from the equation itself, and therefore, quantum mechanics provides a physical basis of the AS and AW.

Paradox of Tunneling When tunneling through a barrier, the wave packet is distorted, and its center is moving backward. Therefore, the center of the transmitted packet leaves the barrier before the center of the incident packet has arrived. This paradox has given birth to many alternative approaches, and time of tunneling is still an open question. The paradox is the same as for classical damping waves, however, and the conflict disappears if we associate the wave packet with an ensemble of particles instead of a single particle.

Nonlinear Waves Using the frequency-separation procedure, we also study nonlinear waves. The most noticeable effect is that *velocity modulation* appears from higher harmonics. In turn, velocity modulation results in *frequency modulation* of waves. On the other hand, frequency modulation can be compensated in a nonlinear dispersive medium. Then *solitary waves* travel without distortions. This amazing nonlinear effect can easily be explained with the AW.

The abundance of the problems confirms that the global AS is important for physics and engineering. Each chapter contains a "Supplementary Problems" section, which may be burdensome in a regular text. Many comments, reassertions, and explanations are included in these problems. However, though interesting or elegant, they may be skipped in the first reading.

Reference

[1] Wainstein, L. A., and D. E. Vakman, *Frequency Separation in Theory of Oscillations and Waves*, Moscow: Nauka-Press, 1983 (in Russian).

2

The Analytic Signal and Its Properties

In this chapter, we consider a method used throughout the book for defining the amplitudes and frequencies of signals and waves. The method is based on the *Hilbert transform* (HT) and the *analytic signal* (AS) introduced by Gabor [1] in 1946. The AS has been studied and applied in many papers and books [2–8]. The HT is also considered in recent books [9,10]. In this chapter, we discuss basic mathematical properties of the AS and HT and their applications to typical signals. More physical and technical implementations are considered in the following chapters.

2.1 Amplitude—Phase Ambiguity

The amplitude and frequency of a real signal

$$u(t) = a(t)\cos\phi(t) \tag{2.1}$$

are ambiguous since (2.1) is an equation with two unknowns—a and ϕ. Therefore, the amplitude and phase cannot be explicitly defined from the given function $u(t)$ or from the oscillogram of a signal. On the other hand, the *complex* signal

$$w(t) = u(t) + iv(t) = a(t)e^{i\phi(t)} \tag{2.2}$$

formed by adding an imaginary part $v(t)$ to the real observed signal $u(t)$ defines the *amplitude, phase, and frequency* (APF) unambiguously:

$$a(t) = \sqrt{u^2(t) + v^2(t)} = |w(t)| \tag{2.3}$$

$$\phi(t) = \arctan\left\{\frac{v(t)}{u(t)}\right\} = \mathrm{Im}\{\ln w(t)\} \tag{2.4}$$

$$\omega(t) = \frac{d\phi}{dt} = \frac{d}{dt}\arctan\left\{\frac{v}{u}\right\} = \frac{1}{1+\frac{v^2}{u^2}}\frac{d}{dt}\left\{\frac{v}{u}\right\} = \frac{v'u - u'v}{u^2+v^2} \tag{2.5}$$

$$= \frac{d}{dt}\mathrm{Im}(\ln w) = \mathrm{Im}\left\{\frac{w'(t)}{w(t)}\right\} \tag{2.6}$$

where we have used both forms (2.4) for the phase ϕ.

However, the imaginary part $v(t)$ cannot be observed in reality. We can only assume that it is somehow linked with the observed signal. Symbolically we write

$$v(t) = \mathcal{H}[u(t)] \tag{2.7}$$

and any operator $\mathcal{H}$ generates the specific APF. In Figure 2.1, the hypothetical measuring device for the APF contains an arbitrary operator $\mathcal{H}$ inside and illustrates the transformations (2.3) to (2.5). There are an infinite number of possibilities for $\mathcal{H}$. Hence, to define the APF, *we must know how real signals $u(t)$ are transformed into conjugated signals $v(t)$*.

It may be argued that the amplitude and frequency are known from the envelope of maximums and zero-crossings in the oscillogram. Then imaginary signals and operators are needless. This reasonable approach is effective for narrowband signals. However, taking the $a(t)$ and $\phi(t)$ from the envelope and zero-crossings, we can construct the imaginary signal as $v(t) = a(t)\sin\phi(t)$. This results in the APF accordingly to (2.3) through (2.6). So, with the envelope and zero-crossings, we employ one of the possible relations between $u(t)$ and $v(t)$.

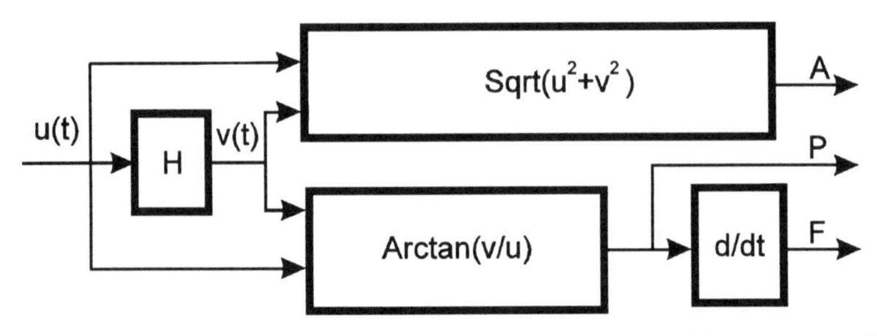

Figure 2.1 Hypothetical measuring device for the APF involving the arbitrary operator H.

2.2 Definition of the Analytic Signal

Since the 19th century, we represent sinusoidal signals $a\cos(\omega t+\Phi)$ of constant a, ω, and Φ as complex exponents $ae^{i(\omega t+\Phi)}$ (or rotating vectors) adding the imaginary part $a\sin(\omega t+\Phi)$ shifted by $-\pi/2$.

The AS is a generalization of this approach. A real signal $u(t)$ of a spectrum

$$U(\omega)=\int_{-\infty}^{\infty} u(t)e^{-i\omega t}\,dt=\int_{-\infty}^{\infty} u(t)(\cos\omega t-i\sin\omega t)\,dt=U_c(\omega)-iU_s(\omega) \tag{2.8}$$

can be written by either of Fourier integrals:

$$u(t)=\frac{1}{2\pi}\int_{-\infty}^{\infty} U(\omega)e^{i\omega t}\,d\omega \tag{2.9}$$

or, since $U_c(\omega)$ is an even and $U_s(\omega)$ is an odd function of ω,

$$u(t)=\frac{1}{\pi}\int_{0}^{\infty}[U_c(\omega)\cos\omega t+U_s(\omega)\sin\omega t]\,d\omega \tag{2.10}$$

Following Gabor, we now create the imaginary part of the AS shifting all phases in (2.10) by $-\pi/2$. Then we come to the Hilbert-conjugated signal as follows:

$$v(t)=\mathcal{H}[u]=\frac{1}{\pi}\int_{0}^{\infty}[U_c(\omega)\sin\omega t-U_s(\omega)\cos\omega t]\,d\omega \tag{2.11}$$

where $\mathcal{H}$ denotes HT. Further, using (2.10) and (2.11), we construct the AS:

$$\begin{aligned} w(t)&=u(t)+iv(t)\\ &=\frac{1}{\pi}\int_{0}^{\infty}[U_c(\omega)(\cos\omega t+i\sin\omega t)+U_s(\omega)(\sin\omega t-i\cos\omega t)]\,d\omega\\ &=\frac{1}{\pi}\int_{0}^{\infty}[U_c(\omega)-iU_s(\omega)]e^{i\omega t}\,d\omega=\frac{1}{\pi}\int_{0}^{\infty}U(\omega)e^{i\omega t}\,d\omega \end{aligned} \tag{2.12}$$

where (2.8) has also been taken into account. Then, comparing with (2.9), we come to relations between spectra of a real signal and its AS:

$$W(\omega)=\begin{cases}2U(\omega) & \text{for } \omega>0\\ 0 & \text{for } \omega<0\end{cases} \tag{2.13}$$

and between spectra of a signal and its HT:

$$V(\omega) = -i[W(\omega) - U(\omega)] = \begin{cases} -iU(\omega) & \text{for } \omega > 0 \\ iU(\omega) & \text{for } \omega < 0 \end{cases} \tag{2.14}$$

Now we formulate basic properties of the HT and AS.

Property 1: Linearity and Superposition The HT is linear; that is,

$$\mathcal{H}\left[\sum_n c_n u_n\right] = \sum_n c_n \mathcal{H}[u_n]$$

where c_n are arbitrary numbers and u_n are arbitrary functions for which HT is defined.

Property 2: Harmonic Correspondence For a signal $u(t) = a\cos(\omega t + \Phi)$ of constant $a > 0$, $\omega > 0$, and Φ, the HT is $v(t) = a\sin(\omega t + \Phi)$, and the AS $w(t) = ae^{i(\omega t+\Phi)}$ produces ordinary APF of all harmonic processes. This is easily verified if we write $\mathcal{H}[a\cos(\omega t + \Phi)] = a\cos(\omega t + \Phi - \pi/2)$ and expand.

The HT is the only operator that meets properties 1 and 2. Really, because of linearity, $v(t)$ can be written as a Fourier integral, and because of harmonic correspondence, this integral takes the form of (2.11) if a real signal is given as in (2.10). Thus, properties 1 and 2 are necessary and sufficient: the APF are accordant to the HT and AS if these properties are met, and vice versa.

Property 3: Positive Definiteness of Spectrum A complex function $w(t)$ is the AS if and only if its Fourier spectrum $W(\omega)$ equals zero at $\omega < 0$. We also note that for signals of finite energy, the integral (2.12) over *positive* ω converges at any complex t on the upper half-plane $\operatorname{Im} t > 0$. So, $w(t)$ is an analytic function on this half-plane that explains the term "analytic signal."

We have considered the HT and AS in a frequency domain where the HT shifts the phases by $-\pi/2$. For coming into a time domain, let us interpret the HT as filtering. The frequency characteristic of a filter associated with HT results from (2.14):

$$H(\omega) = \frac{V(\omega)}{U(\omega)} = \begin{cases} -i & \text{for } \omega > 0 \\ i & \text{for } \omega < 0 \end{cases} \tag{2.15}$$

Therefore, its impulse response is as follows:

$$h(t) = \frac{1}{2\pi}\int_{-\infty}^{\infty} H(\omega)e^{i\omega t}\,d\omega = \frac{-i}{2\pi}\left\{\int_0^{\infty} - \int_{-\infty}^{0}\right\} e^{i\omega t}\,d\omega$$

$$= \frac{1}{\pi}\int_0^{\infty} \sin\omega t\,d\omega = -\frac{\cos\omega t|_0^{\infty}}{\pi t} = \frac{1}{\pi t} \tag{2.16}$$

Finally, in a time domain, the HT is the convolution

$$v(t) = \int_{-\infty}^{\infty} u(s)h(t-s)\,ds = \frac{1}{\pi}\int_{-\infty}^{\infty}\frac{u(s)}{t-s}\,ds \tag{2.17}$$

where the Caushy principal value of integral is implied (see Problem 2.2). This formula immediately leads to the following property.

Property 4: Time Stationarity If a signal $u(t)$ is replaced by $u(t-t_0)$, where t_0 is arbitrary delay, then $v(t)$ and $w(t)$ are replaced by $v(t-t_0)$ and $w(t-t_0)$, respectively. This means also that HT is commutative with other stationary transformations (e.g., differentiation):

$$\mathcal{H}\left[\frac{du}{dt}\right] = \frac{d}{dt}\mathcal{H}[u] \qquad \mathcal{H}\left[\frac{d^2u}{dt^2}\right] = \frac{d^2}{dt^2}\mathcal{H}[u]$$

and so forth.

The transformations are commutative since the associated frequency characteristics are multiplied. In fact, in a frequency domain, $i\omega$ is the operator of differentiation, and according to (2.14), $-i\ \text{sgn}(\omega)$ is the operator of HT. Here $\text{sgn}(\omega)$ denotes the signum function given by

$$\text{sgn}(\omega) = \begin{cases} 1 & \text{for } \omega > 0 \\ 0 & \text{for } \omega = 0 \\ -1 & \text{for } \omega < 0 \end{cases}$$

Hence the product $\{i\omega\}\{-i\ \text{sgn}(\omega)\}U(\omega)$ defines the spectrum of the derivative of the HT of $u(t)$, and the same product in another order $\{-i\ \text{sgn}(\omega)\}\{i\omega\}U(\omega)$ defines the spectrum of the HT of the derivative of $u(t)$.

2.3 Bandpass Signals and Slow-Time Functions

Bandpass signals (BPSs) are employed in radio communications, and their bands are much less than carrier frequencies for neighboring channels to exist. Now we show that the AS defines the amplitude and frequency of BPSs in the same way as for harmonic signals. We also prove Bedrosian's theorem [15], which generalizes averaging and similar approximate methods.

Bandpass (narrowband) signals are given by

$$u(t) = a(t)\cos[\omega_0 t + \Phi(t)] = x(t)\cos\omega_0 t - y(t)\sin\omega_0 t \tag{2.18}$$

where the quadratures $x(t) = a(t)\cos\Phi(t)$ and $y(t) = a(t)\sin\Phi(t)$ are band-limited to $\omega < \omega_0$. Slow amplitude and phase of the BPS are defined as

$$a(t) = \sqrt{x^2(t) + y^2(t)} \qquad \Phi(t) = \arctan\left\{\frac{y(t)}{x(t)}\right\} \tag{2.19}$$

So, we do not use the AS or the complex form (2.2) but associate the APF with low-frequency quadratures. However, the AS gives the same APF.

2.3.1 Bedrosian's Theorem

Applying the HT to (2.18), we first consider a special case:

$$u(t) = \cos\omega t \cos\omega_0 t = \frac{\cos(\omega_0 - \omega)t + \cos(\omega_0 + \omega)t}{2}$$

where $\omega < \omega_0$. Then, using properties 1 and 2, we find

$$\begin{aligned} \mathcal{H}[u(t)] &= \frac{\sin(\omega_0 - \omega)t + \sin(\omega_0 + \omega)t}{2} \\ &= \cos\omega t \cdot \sin\omega_0 t = \cos\omega t \cdot \mathcal{H}[\cos\omega_0 t] \end{aligned}$$

So, the low-frequency factor $\cos\omega t$ can be taken out of the HT. We can see that the same applies to $\sin\omega t$ or $\sin\omega_0 t$.

Band-limited quadratures in (2.18) consist of harmonic components with $\omega < \omega_0$. Each component can be taken out of the HT, and because of linearity, the same is true for the quadratures $x(t)$ and $y(t)$ themselves. So, we find

$$\begin{aligned} v(t) &= \mathcal{H}[x(t)\cos\omega_0 t - y(t)\sin\omega_0 t] = x(t)\mathcal{H}[\cos\omega_0 t] - y(t)\mathcal{H}[\sin\omega_0 t] \\ &= x(t)\cos(\omega_0 t - \pi/2) - y(t)\sin(\omega_0 t - \pi/2) \\ &= x(t)\sin\omega_0 t + y(t)\cos\omega_0 t \end{aligned} \tag{2.20}$$

$$\begin{aligned} w(t) &= u + iv = x(t)[\cos\omega_0 t + i\sin\omega_0 t] - y(t)[\sin\omega_0 t - i\cos\omega_0 t] \\ &= [x(t) + iy(t)]e^{i\omega_0 t} \end{aligned} \tag{2.21}$$

where property 2 has been consulted again. Comparing with (2.19), we see that the AS defines the same APF of bandpass signals. Generalizing (2.20), we come to the following property.

Property 5: Bedrosian's Theorem If the product $l(t) \cdot h(t)$ consists of a low-frequency factor $l(t)$ and a high-frequency factor $h(t)$ of *nonoverlapping spectra,* then a low-frequency factor can be taken out of the HT:

$$\mathcal{H}[l(t) \cdot h(t)] = l(t) \cdot \mathcal{H}[h(t)] \tag{2.22}$$

To prove the theorem, we note that $h(t)$ consists of high-frequency harmonic components, and (2.20) is applicable to each of them. Spectral nonoverlapping means that there exists the boundary frequency ω_0 such that all frequencies in $l(t)$ are lower and all frequencies in $h(t)$ are higher than ω_0.

So, we *freeze* a low-frequency modulating factor $l(t)$ in the HT as if it were a constant factor. When averaging, we also freeze slow modulation and take it out of integrals. However, in contrast to averaging, transformation (2.22) is exact, and condition of spectral nonoverlapping is less restrictive than slowness. If, for example, the carrier frequency of 1 MHz is modulated with 999 kHz, Bedrosian's theorem is applicable, though modulation is not slow with respect to the carrier.

2.3.2 Asymptotic Approach and FM Signals

Another signal model is also used for slow modulation. Assuming $\varepsilon \ll 1$, we rewrite (2.18) as follows (without a loss of generality we set $\omega_0 = 1$):

$$u(t) = a(\varepsilon t)\cos[t + \Phi(\varepsilon t)] = x(\varepsilon t)\cos t - y(\varepsilon t)\sin t \tag{2.23}$$

Though the quadratures depend on slow-time εt, their low-frequency spectra may leak behind the carrier $\omega_0 = 1$. Nevertheless, Bedrosian's theorem can be applied *asymptotically* in the following sense.

According to the well-known property (see also Problem 2.9), if a function of εt has r derivatives, its spectrum is decreasing at high frequencies as $(\varepsilon/\omega)^{r+1}$, and for the quadratures in (2.23) we have

$$X(\omega) \sim \left\{\frac{\varepsilon}{\omega}\right\}^{r+1} \qquad Y(\omega) \sim \left\{\frac{\varepsilon}{\omega}\right\}^{r+1} \tag{2.24}$$

Neglecting the spectra at $\omega \geq 1$, we allow an error of the order ε^{r+1}, and Bedrosian's theorem is applicable with the same small error. For the smooth signal in Problem 2.10 and for frequency modulation discussed below, $r=\infty$, and the error is exponentially small.

Amplitude modulation is typically narrowband, and for the AM signal $u(t)=a(t)\cos t$, Bedrosian's theorem is strictly applicable if $a(t)$ is band-limited to $\omega<1$. Therefore, the AS $w(t)=a(t)e^{it}$ defines the expected amplitude of AM signals: $|w(t)|=a(t)$. For frequency modulation, however, the context is more sophisticated. Even for one-tone modulation $u(t)=\cos(t+m\sin\varepsilon t)$ with $\varepsilon\ll 1$, spectra of quadratures $x(t)=\cos(m\sin\varepsilon t)$ and $y(t)=\sin(m\sin\varepsilon t)$ leak behind the carrier, and the complex function

$$w(t) = e^{i(t+m\sin\varepsilon t)} \tag{2.25}$$

is *not* the AS. Moreover, it was rigorously shown that, in a strict sense, the AS represents a very few FM signals far from the practice [16].

Nevertheless, the spectrum of (2.25) is small at negative frequencies. Reducing it to Bessel functions:

$$J_n(m) = \frac{1}{2\pi}\int_{-\pi}^{\pi} e^{i(m\sin x-nx)}\,dx$$

we can show that, outside the bounds $b_{1,2}=1\pm m\varepsilon$ of frequency deviation, the spectrum decreases exponentially (Figure 2.2). All the more, this is true for $\omega<0$. Neglecting the spectrum at negative frequencies, we come to the AS, and therefore, the function (2.25) differs from the AS by a small error. This asymptotic approach (for finite or infinite order r) is very useful in nonlinear oscillation theory (Chapter 6). It is also typical for radio-engineering where modulation is separated with filters (Chapter 4).

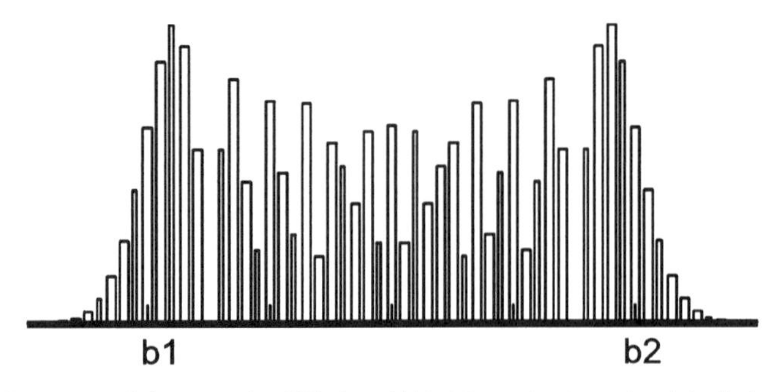

Figure 2.2 Spectrum of the complex FM signal (*b*1, *b*2 are the bounds of deviation).

2.3.3 Relation to Digital Signal Processing

For digital processing, continuous signals are sampled in the *analog-digital converter* (ADC), and the Nyquist condition

$$\omega_{\max} < \frac{\pi}{T}$$

is to be met. Here $\omega_{\max}$ is the maximal frequency in a spectrum, and T is the sampling period. So, the signals must be band-limited.

Therefore, Bedrosian's theorem is applicable, and the APF should be accordant to the AS. This point is often ignored in digital signal processing, and alternative APFs are suggested. Some of them are discussed in Section 3.5.3. In Chapter 5, we also discuss the ADC and Nyquist condition in relation to frequency measurement.

2.4 Complex Filtering and the AS

If a real signal $u(t)$ passes through a filter $K(\omega)$, the output spectrum is the product $U(\omega) \cdot K(\omega)$. According to (2.13), to form the *output* AS, we must restrict this product to positive frequencies. This can be done in three ways: by restricting either $U(\omega)$, $K(\omega)$, or both. Therefore, we have

$$W(\omega) = \begin{cases} 2 \cdot \hat{U}(\omega) \cdot K(\omega) \\ 2 \cdot U(\omega) \cdot \hat{K}(\omega) \\ 2 \cdot \hat{U}(\omega) \cdot \hat{K}(\omega) \end{cases} \tag{2.26}$$

where a "hat" above denotes the restricted spectrum. In a time domain, each product in (2.26) results in a convolution of time functions, and the restricted spectra $2\hat{U}(\omega)$ and $2\hat{K}(\omega)$ correspond to the associated AS. Therefore, we obtain three equivalent forms for the output AS:

$$w(t) = \begin{cases} \displaystyle\int_{-\infty}^{\infty} \hat{u}(s)k(t-s)\,ds \\ \displaystyle\int_{-\infty}^{\infty} u(s)\hat{k}(t-s)\,ds \\ \displaystyle\frac{1}{2}\int_{-\infty}^{\infty} \hat{u}(s)\hat{k}(t-s)\,ds \end{cases} \tag{2.27}$$

where now the hat denotes analytic signals $\hat{u}(t) = u + i\mathcal{H}[u]$ and $\hat{k}(t) = k + i\mathcal{H}[k]$.

For signal processing, the second line in (2.27) is of special interest because it avoids the digital or analog HT for real signals. If the envelope $b(t)$ of the filter's impulse response $b(t)\cos\omega_0 t$ is band-limited to $\omega < \omega_0$, the complex impulse response $b(t)e^{i\omega_0 t}$ is the AS restricted to $\omega > 0$. Convolving a real input signal with this complex response, we suppress noise and produce the output AS as well. This procedure is known in radio-engineering as *quadrature processing* when a signal is passed through two filters with responses $b(t)\cos\omega_0 t$ and $b(t)\sin\omega_0 t$ [11]. The same is achieved in the synchronous detector (see Chapter 4).

2.5 Mean Frequency and Bandwidth

Now we consider the mean frequency and bandwidth of a signal. Spectra of real signals are symmetric at positive and negative frequencies. Therefore, within a framework of real signals, the mean spectral frequency is zero, whatever the signal. Considering positive spectral frequencies only, we come to the AS and define the mean frequency and bandwidth as follows. Let us introduce the first and second spectral moments plus the bandwidth $\Delta\omega$ by equations:

$$\overline{\overline{\omega}} = \frac{\int_0^\infty \omega |W(\omega)|^2\, d\omega}{\int_0^\infty |W(\omega)|^2\, d\omega} \qquad \overline{\overline{\omega^2}} = \frac{\int_0^\infty \omega^2 |W(\omega)|^2\, d\omega}{\int_0^\infty |W(\omega)|^2\, d\omega} \tag{2.28}$$

$$\Delta\omega^2 = \overline{\overline{\omega^2}} - \overline{\overline{\omega}}^2 = \overline{\overline{(\omega - \overline{\overline{\omega}})^2}}$$

Here a double overbar denotes averaging in frequency, and $\overline{\overline{\omega}}$ is the mean spectral frequency of a signal.

Further, for any two signals w_1, w_2, and their spectra W_1, W_2, Parseval's equality is valid:

$$\int_{-\infty}^{\infty} w_1(t) w_2^*(t)\, dt = \frac{1}{2\pi}\int_{-\infty}^{\infty} W_1(\omega)\, W_2^*(\omega)\, d\omega \tag{2.29}$$

and for the AS, the limits in the right-hand integral are replaced by 0, ∞. Applying this equality to the Fourier pairs $w(t) = ae^{i\phi} \leftrightarrow W(\omega)$ and $w'(t) =$

$(a' + ia\phi')e^{i\phi} \leftrightarrow i\omega W(\omega)$, we find

$$\begin{aligned}
\frac{1}{2\pi}\int_0^{\infty} \omega|W(\omega)|^2\,d\omega &= -\frac{i}{2\pi}\int_0^{\infty} i\omega W W^*\,d\omega \\
&= -i\int_{-\infty}^{\infty} [a' + i\phi' a]e^{i\phi} a e^{-i\phi}\,dt \\
&= -\frac{i}{2}a^2(t)\Big|_{-\infty}^{\infty} + \int_{-\infty}^{\infty} \omega(t)a^2(t)\,dt \\
&= \int_{-\infty}^{\infty} \omega(t)a^2(t)\,dt
\end{aligned}$$

$$\begin{aligned}
\frac{1}{2\pi}\int_0^{\infty} \omega^2|W(\omega)|^2\,d\omega &= \frac{1}{2\pi}\int_0^{\infty} |i\omega W(\omega)|^2\,d\omega = \int_{-\infty}^{\infty} |a' + i\phi' a|^2\,dt \\
&= \int_{-\infty}^{\infty} [a'(t)^2 + \omega^2(t)a^2(t)]\,dt
\end{aligned}$$

and finally, formulas (2.28) take the forms

$$\begin{aligned}
\overline{\overline{\omega}} &= \frac{\int_{-\infty}^{\infty} \omega(t)a(t)^2\,dt}{\int_{-\infty}^{\infty} a(t)^2\,dt} = \overline{\omega(t)} \\
\overline{\overline{\omega^2}} &= \frac{\int_{-\infty}^{\infty}\left(\frac{a'(t)^2}{a(t)^2} + \omega(t)^2\right)a(t)^2\,dt}{\int_{-\infty}^{\infty} a(t)^2\,dt} = \overline{\left(\frac{a'(t)}{a(t)}\right)^2} + \overline{\omega(t)^2} \\
\Delta\omega^2 &= \overline{\left(\frac{a'(t)}{a(t)}\right)^2} + \overline{\omega(t)^2} - \overline{\omega(t)}^2 = \overline{\left(\frac{a'(t)}{a(t)}\right)^2} + \overline{(\omega(t) - \overline{\omega(t)})^2}
\end{aligned} \tag{2.30}$$

Here a single overbar denotes averaging in time with $a(t)^2$ as a weighting function.

We have come to important relations. The mean spectral frequency $\overline{\overline{\omega}}$ equals the instantaneous frequency $\overline{\omega(t)}$ *averaged in time*, and the bandwidth $\Delta\omega$ depends on amplitude and frequency variations also averaged in time. Moreover, amplitude and frequency components are summed in quadratures without interaction. These relations originated by Fink [12] are employed in many works; for example, [13,14]. So, the AS leads to the reasonable mean frequency and bandwidth of a signal. Interrelations between spectral and instantaneous frequencies will be detailed in Chapter 3.

2.6 Supplementary Problems

Problem 2.1

Find the inverse HT defining $u(t)$ from $v(t)$.

Solution The HT shifts the phases by $-\pi/2$. Applying it twice, we shift the phases by $-\pi$ and change the sign of a function. Therefore, $\mathcal{H}[v] = \mathcal{H}[\mathcal{H}[u]] = -u$, and finally $u(t) = \mathcal{H}^{-1}[v(t)] = -\mathcal{H}[v(t)] = -\frac{1}{\pi}\int_{-\infty}^{\infty}\frac{v(s)}{t-s}\,ds$

Problem 2.2

For a complex t, the integral (2.12) defines an analytic function regular on the upper half-plane $\mathrm{Im}\, t > 0$. Show the converse: If an analytic function $w(t)$ is regular on the upper half-plane and vanishes at $|t| \to \infty$, then $w(t)$ is the AS in the real axis.

Solution Due to Caushy formula (see Problem 2.13), we have

$$w(t) = \frac{1}{2\pi i}\oint_C \frac{w(s)}{s-t}\,ds$$

where a contour C surrounds the point t on the upper half-plane. For a real t, the contour can be chosen as in Figure 2.3, where a small semicircle of a radius ε encircles the point t.

Because of regularity, we can widen the large semicircle to include the whole half-plane. Because w vanishes for $|t| \to \infty$, the contribution from the

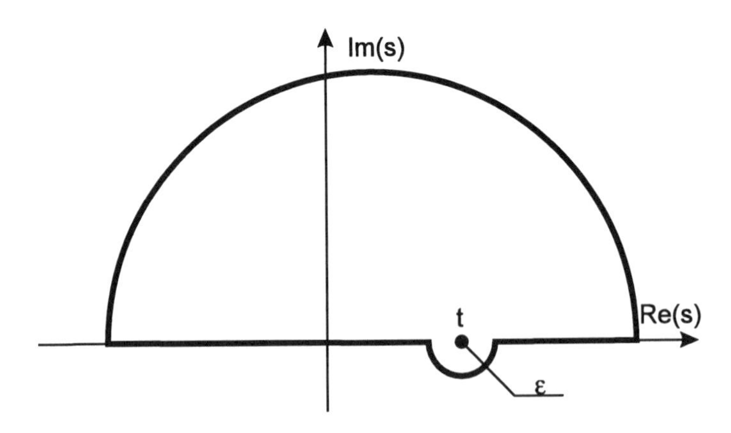

Figure 2.3 Path of integration in Caushy formula.

large semicircle is zero, and we have

$$w(t) = \frac{w(t)}{2} + \frac{1}{2\pi i}\left\{\int_{-\infty}^{t-\varepsilon} + \int_{t+\varepsilon}^{\infty}\right\}\frac{w(s)}{s-t}\,ds$$

Here, the first addend is the contribution from the small semicircle, and for $\varepsilon \to 0$, the sum of integrals is the integral in the Caushy sense. So, we obtain

$$w(t) = \frac{1}{\pi i}\int_{-\infty}^{\infty}\frac{w(s)}{s-t}\,ds$$

Setting $w = u + iv$, we separate the real and imaginary parts and finally come to the HT (2.17):

$$v(t) = \frac{1}{\pi}\int_{-\infty}^{\infty}\frac{u(s)}{t-s}\,ds, \qquad u(t) = -\frac{1}{\pi}\int_{-\infty}^{\infty}\frac{v(s)}{t-s}\,ds$$

Therefore, $w(t)$ is the AS. We also obtained the result of Problem 2.1 in another way.

Problem 2.3

Show that a product of two ASs is also AS and interpret a set of ASs as a multiplicative group. Show that a constant is the AS, and therefore the second condition in Problem 2.2 can be diminished. (See also Problems 2.4 and 2.13.)

Solution The product $w = w_1 w_2$ meets the conditions of Problem 2.2. Indeed, w is regular on the upper half-plane because of regularity of w_1 and w_2, and $w \to 0$ for $|t| \to \infty$ because w_1 and w_2 do so.

A sum of two ASs is the AS also (Property 1). Therefore, the set of AS is a multiplicative group for the common addition and multiplication. However, the multiplicative unit of the group must be included, so that the constant $w(t) = 1$ must be the AS. This is true because, for $u(t) = 1$, we have

$$v(t) = \frac{1}{\pi}\int_{-\infty}^{\infty}\frac{ds}{t-s} = \frac{1}{\pi}\left\{\int_{-\infty}^{-\varepsilon} + \int_{\varepsilon}^{\infty}\right\}\frac{ds}{s} = \frac{1}{\pi}\left\{\int_{\varepsilon}^{\infty} - \int_{\varepsilon}^{\infty}\right\}\frac{ds}{s} = 0$$

where it was set $t - s = r$ and then renamed $r = s$. Hence $w = u + iv = 1$. The Caushy principal value of integral has been taken into account here.

Problem 2.4

Let $f(w)$ be an entire function of a complex w, and let $w(t)$ be the AS. Show that $f[w(t)]$ is also the AS.

Solution First we set $f(0)=0$. Then $f[w(t)]$ meets the conditions of Problem 2.2. Really, it is regular due to regularity of f and w, and for $|t| \to \infty$, it runs to zero together with the w because of the Taylor series (see Problem 2.13).

$$f(w) = f'(0)w + \frac{f''(0)}{2}w^2 + \cdots$$

In general, $f(w)$ differs from that considered by a constant $f(0)$. According to Problem 2.3, however, a constant is the AS also, and $f(w)$ is the AS as a sum of two ASs.

Problem 2.5

Let the amplitude and phase of a signal $u(t) = a(t)\cos\phi(t)$ be accordant to the AS. Find the HT of the function $u_n(t) = a^n(t)\cos n\phi(t)$ for an integer n.

Solution The function $w = ae^{i\phi}$ is the AS. By virtue of Problem 2.4, $w^n = a^n e^{in\phi}$ is also the AS. Its real part is the u_n, and therefore, its imaginary part is the HT desired: $v_n(t) = a^n(t)\sin n\phi(t)$. This transformation of the AS is typical for frequency multipliers (Chapter 4).

Problem 2.6

Because of the causality principle, the impulse response $k(t)$ of a causal filter is zero at $t < 0$. Show that real and imaginary parts of its transfer function $K(\omega)$ are interrelated with the HT.

Solution The transfer function is the Fourier transform of the impulse response

$$K(\omega) = U(\omega) + iV(\omega) = \int_0^{\infty} k(t)e^{-i\omega t}\,dt$$

where the integral is taken over positive t because of causality and analogous to the integral (2.12) for the AS. According to Property 3, the Fourier transform $K(\omega)$ of a function $k(t)$ that is zero at $t < 0$ has the same properties as the AS. Therefore, its real and imaginary parts are interrelated with the HTs:

$$V(\omega) = -\frac{1}{\pi}\int_{-\infty}^{\infty} \frac{U(\zeta)}{\omega - \zeta}\,d\zeta \qquad U(\omega) = \frac{1}{\pi}\int_{-\infty}^{\infty} \frac{V(\zeta)}{\omega - \zeta}\,d\zeta$$

Problem 2.7

Let $u(t)$ be a *periodic* function of the period T. Show that its HT takes the form

$$v(t) = \frac{1}{T}\int_0^T u(s)\cot\left(\pi\frac{t-s}{T}\right)ds \qquad -\infty < t < \infty \tag{2.31}$$

Use the following expansion for cotangent:

$$\cot \pi x = \frac{1}{\pi}\sum_{n=-\infty}^{\infty}\frac{1}{x-n}$$

Also, applying (2.31) to a function given on the finite interval $(0,\ T)$, interpret this transformation in a frequency domain.

Solution Because of the periodicity of $u(t)$, we find from (2.17)

$$\begin{aligned} v(t) &= \frac{1}{\pi}\int_{-\infty}^{\infty}\frac{u(s)}{t-s}ds = \sum_{n=-\infty}^{\infty}\frac{1}{\pi}\int_{nT}^{(n+1)T}\frac{u(s)}{t-s}ds \\ &= \sum_{n=-\infty}^{\infty}\frac{1}{\pi}\int_0^T\frac{u(s+nT)}{t-s-nT}ds = \sum_{n=-\infty}^{\infty}\frac{1}{\pi}\int_0^T\frac{u(s)}{t-s-nT}ds \end{aligned}$$

Then, changing the order of integration and summation and using the above expansion for cotangent, we come to (2.31).

When applying to a function given on the finite interval $(0,\ T)$, we extend the function periodically over the whole axis t. Therefore, the HT shifts phases by $-\pi/2$ in the Fourier series instead of the Fourier integral.

Problem 2.8

Show that the HT of a bandpass signal $u(t)$ commutates with multiplication by t^n (and therefore, by a polynomial of t):

$$\mathcal{H}[t^n \cdot u(t)] = t^n \cdot \mathcal{H}[u(t)] \tag{2.32}$$

Interpret this relation with Bedrosian's theorem.

Solution Since $\frac{s}{t-s} = \frac{t}{t-s} - 1$, we find

$$\begin{aligned} \mathcal{H}[tu] &= \frac{1}{\pi}\int_{-\infty}^{\infty}\frac{s\,u(s)}{t-s}ds \\ &= \frac{t}{\pi}\int_{-\infty}^{\infty}\frac{u(s)}{t-s}ds - \frac{1}{\pi}\int_{-\infty}^{\infty}u(s)\,ds = t\mathcal{H}[u] - \frac{U(0)}{\pi} \end{aligned}$$

where $U(\omega)$ is the spectrum of $u(t)$ [see (2.8) with $t=s$]. For multiplication by t^n, we obtain by induction

$$\mathcal{H}[t^n u] = t^n \mathcal{H}[u] - \frac{1}{\pi}\sum_{k=0}^{n-1} i^k\, t^{n-1-k}\, \frac{d^k U}{d\omega^k}(0)$$

If $u(t)$ is a bandpass signal, $U(\omega)$ and its derivatives are zero at $\omega = 0$. Then we come to (2.32).

Differentiating the δ-function given by

$$2\pi\,\delta(\omega) = \int_{-\infty}^{\infty} e^{-i\omega t}\, dt$$

we have

$$2\pi\,\delta^{(n)}(\omega) = \int_{-\infty}^{\infty} (-it)^n e^{-i\omega t}\, dt.$$

Therefore, the n-th derivative $\delta^{(n)}(\omega)$ defines the spectrum of $(-it)^n$. Since $\delta^{(n)}(\omega)$ is concentrated at $\omega = 0$, spectra of t^n and a bandpass signal are nonoverlapping. Hence, because of Bedrosian's theorem, t^n can be taken out of the HT. Then we again obtain (2.32).

Problem 2.9

Let $x(t)$ and its derivatives behave well at $t \to \pm\infty$, and let the r-th derivative be discontinuous at $t = c$ so that $x^{(r)}(c+0) - x^{(r)}(c-0) = A$. Then the $(r+1)$-th derivative can be written as

$$x^{(r+1)}(t) = A\delta(t-c) + f(t)$$

where $f(t)$ is continuous. Show that the spectrum of $x(t)$ is estimated as

$$X(\omega) \sim \frac{1}{\omega^{r+1}} \quad \text{for } \omega \to \infty \tag{2.33}$$

Show also that if x depends on a slow time εt, the estimate (2.24) is true.

Solution We first set $r = 0$. Then the above conditions relate to $x(t)$ and its first derivative $x'(t)$. Therefore, taking the integral by parts, we have

$$\begin{aligned} X(\omega) &= \int_{-\infty}^{\infty} x(t)e^{-i\omega t}\, dt = \frac{1}{i\omega}\int_{-\infty}^{\infty} x'(t)e^{-i\omega t}\, dt \\ &= \frac{1}{i\omega}\int_{-\infty}^{\infty} [A\delta(t-c) + f(t)]e^{-i\omega t}\, dt = \frac{1}{i\omega}[Ae^{-i\omega c} + F(\omega)] \end{aligned}$$

where the spectrum $F(\omega)$ of a continuous function $f(t)$ approaches zero at $\omega \to \infty$. So, for $r = 0$, the estimate (2.33) is true. For $r > 0$, we have obtained the spectrum of the r-th derivative, so that

$$(i\omega)^r X(\omega) = \frac{1}{i\omega}[Ae^{-i\omega c} + F(\omega)]$$

and (2.33) is true generally. If $x(t)$ is replaced by $x(\varepsilon t)$, then for $\varepsilon > 0$, its spectrum is proportional to $X(\omega/\varepsilon)$, and (2.33) turns into (2.24).

Problem 2.10

We often replace the real signal $u(t) = \exp(-\varepsilon^2 t^2)\cos\omega_0 t$ with the *analytic* (entire) function $w(t) = \exp(-\varepsilon^2 t^2 + i\omega_0 t)$. Show that this analytic function is *not* AS but *approximates* it in the asymptotic sense of Section 2.3.2. Show also that the order of approximation is infinite, and the error is exponentially small. Consider relation to Problem 2.2.

Solution For any finite ε and ω_0, the spectrum of $w(t)$ leaks behind $\omega = 0$. Therefore, $w(t)$ is not the AS. Also, $w(t) \to \infty$ at $t \to i\infty$, which violates the condition of Problem 2.2.

However, the asymptotic approach of Section 2.3.2 can be applied, and the function $x(t) = \exp(-\varepsilon^2 t^2)$ is infinitely differentiable. Therefore, $r = \infty$ in (2.24), and the error of approximation runs to zero faster than any finite power of ε. Such an error is considered exponentially small.

Problem 2.11

The operator $\mathcal{A}$ converting real signals into ASs can be written as $\mathcal{A} = \mathcal{I} + i\mathcal{H}$ where $\mathcal{I}$ is the unit operator and $\mathcal{H}$ is the HT. Show that operator $\mathcal{P} = \mathcal{A}/2$ yields $\mathcal{P}^2 = \mathcal{P}$ and, hence, is a projecting operator [11, 17, 22, 42]. Interpret this operator in a frequency domain.

Solution According to Problem 2.1, $\mathcal{H}^2 = -\mathcal{I}$, and besides $\mathcal{I}^2 = \mathcal{I}$. Therefore

$$\mathcal{P}^2 = \frac{1}{4}[\mathcal{I} + i\mathcal{H}]^2 = \frac{1}{4}[\mathcal{I}^2 + 2i\mathcal{I}\mathcal{H} - \mathcal{H}^2] = \frac{1}{2}[\mathcal{I} + i\mathcal{H}] = \mathcal{P}$$

and $\mathcal{P}$ is a projecting operator. It restricts spectra according to (2.13):

$$W_p(\omega) = \begin{cases} U(\omega) & \text{for } \omega > 0 \\ 0 & \text{for } \omega < 0 \end{cases}$$

So, the operator $\mathcal{P}$ projects real signals onto the subspace of functions with positively defined spectra. The (orthogonal) projection provides the best approximation, and the condition $\mathcal{P}^2 = \mathcal{P}$ shows that the repeated projection does not change a function. Really, the repeated band-limiting leaves the spectrum unchanged if it has been band-limited.

Problem 2.12

Show that the AS amplitude $a(t) = |w(t)|$ defines the energy of a signal as follows:

$$E = \frac{1}{2} \int_{-\infty}^{\infty} a^2(t)\, dt \tag{2.34}$$

Solution Using the spectral relation (2.13) and Parseval's equality (2.29), we come to (2.34):

$$\begin{aligned} E &= \int_{-\infty}^{\infty} u^2(t)\, dt = \frac{1}{2\pi} \int_{-\infty}^{\infty} |U(\omega)|^2\, d\omega = \frac{1}{\pi} \int_{0}^{\infty} |U(\omega)|^2\, d\omega \\ &= \frac{1}{4\pi} \int_{0}^{\infty} |W(\omega)|^2\, d\omega = \frac{1}{2} \int_{-\infty}^{\infty} |w(t)|^2\, dt \end{aligned}$$

Problem 2.13

In this problem, we review some definitions and properties related to functions of a complex variable.

Analytic Functions The function $w(t)$ of a complex variable t is *analytic* (or *regular*) at the point t if its derivative

$$w'(t) = \lim_{\delta \to 0} \frac{w(t+\delta) - w(t)}{\delta}$$

exists in some neighborhood of the t. The function is analytic inside a region if it is analytic at each of its points.

Common rules of differentiation are applicable, and for the AS (2.12) we have

$$w'(t) = \frac{1}{\pi} \int_{0}^{\infty} i\omega U(\omega) e^{i\omega t}\, d\omega$$

This integral converges at any t on the upper half-plane $\mathrm{Im}t > 0$ because, for a positive ω, the exponent $|e^{i\omega t}| = \exp(-\omega \mathrm{Im}t)$ is damping. Therefore, the AS is an analytic function inside the upper half-plane.

Taylor Series If $w(t)$ is analytic inside a circle centered at t_0, its Taylor series

$$w(t) = \sum_{n=0}^{\infty} \frac{w^{(n)}(t_0)}{n!}(t - t_0)^n$$

converges for any t inside the circle, and all derivatives $w^{(n)}(t_0)$ exist.

Entire Functions The function $w(t)$ is *entire* if it is analytic at any finite point on the complex t-plane. Some properties of entire functions are given in Problem 4.2.

Contour Integrals If $w(t)$ is analytic inside a region surrounded by the closed contour C, then (according to the Caushy theorem)

$$\oint_C w(t)\,dt = 0$$

and (according to the Caushy formula)

$$\frac{1}{2\pi i}\oint_C \frac{w(t)}{t - t_0}\,dt = w(t_0)$$

where t_0 is a point inside the region.

Let us prove the Caushy formula. From the Caushy theorem, we can shrink the contour C to a small circle of a radius ε centered at t_0. Then, along the circle, $t - t_0 = \varepsilon e^{i\phi}$ and $dt = \varepsilon e^{i\phi} i d\phi$. Also $w(t) \to w(t_0)$ for $\varepsilon \to 0$, and we find

$$\frac{1}{2\pi i}\oint_C \frac{w(t)}{t - t_0}\,dt = \frac{w(t_0)}{2\pi i}\int_0^{2\pi} \frac{\varepsilon e^{i\phi}}{\varepsilon e^{i\phi}} i d\phi = \frac{w(t_0)}{2\pi}\int_0^{2\pi} d\phi = w(t_0)$$

In Problem 2.2, the integral along a small semicircle is taken in the same way.

References

[1] Gabor, D., "Theory of Communication," *Jour. IEE*, Vol. 93, 1946, pp. 429–457.

[2] Bedrosian, E., "The Analytic Signal Representation of Modulated Waveforms," *Proc. of the IRE*, Vol. 50, 1962, pp. 2071–2076.

[3] Dugundji, J., "Envelopes and Preenvelopes of Real Waveforms," *IRE Trans. Inf. Theory*, Vol. 4, 1958, pp. 53–57.

[4] Oswald, J., "The Theory of Analytic Band-Limited Signals Applied to Carrier Systems," *IRE Trans. Comm. Theory*, Vol. 3, 1956, pp. 244–251.

[5] Picinbono, B., *Principles of Signals and Systems*, London: Artech House, 1988.

[6] Rihaczec, A. W., *Principles of High-Resolution Radar*, Los Altos, CA: Peninsula, 1985.

[7] Urkowitz, H., *Signal Theory and Random Processes*, Dedham, MA: Artech House, 1983.

[8] Ville, J., "Théorie et Applications de la Notion de Signal Analytique," *Cables et Transmissions*, Vol. 2, 1948, pp. 61–74.

[9] Hahn, S. L., *Hilbert Transforms in Signal Processing*, Norwood, MA: Artech House Inc., 1997.

[10] Ponlarikas, A. D., ed., *The Transforms and Applications Handbook*, Boca Raton, FL: CRC Press, 1996.

[11] Frenks, L. E., *Signal Theory*, Englewood Cliffs, NJ: Prentice Hall, 1969.

[12] Fink, L. M., "Relations between the Spectrum and Instantaneous Frequency of a Signal," translated in *Problems Inform. Transm.*, Vol. 2, 1966, pp. 11–21.

[13] Cohen, L., *Time-Frequency Analysis*, Englewood Cliffs, NJ: Prentice Hall, 1995.

[14] Mandel, L., "Interpretation of Instantaneous Frequency," *Amer. Jour. Phys.*, Vol. 42, 1974, pp. 840–846.

[15] Bedrosian, E., "A Product Theorem for Hilbert Transforms," *Proc. of the IEEE*, Vol. 51, 1963, pp. 868–869.

[16] Picinbono, B., "On Instantaneous Amplitude and Phase of Signals," *IEEE Trans. Signal Processing*, Vol. 45, 1997, pp. 552–560.

[17] Lusternik, L. A., V. I. Sobolev, *Elements of Functional Analysis*, Moscow: Nauka-Press, 1968 (in Russian).

3

Local and Global

Amplitude and frequency of a signal are *local* in the sense that they can be varied (modulated) or estimated (received, detected) at each instant. Moreover, in radio broadcasting, nobody knows beforehand what the announcer will say at a given moment. Therefore, the amplitude and frequency modulated *must* be and *are* local since broadcasting does exist.

Nevertheless, amplitude and frequency are not quite local. Modulators and detectors have an averaging time Δt of about few periods of a carrier (intermediate) frequency, and the APF are modulated and detected during this finite time. Usually, the Δt is short with respect to modulation, but the question is whether this limitation is fundamental or technical. In other words, whether we can come to the limit $\Delta t \to 0$ like for the velocity $\frac{\Delta x}{\Delta t}$, or the Δt has to be finite for the APF. If so, how long must it be?

If we use the AS, the context is changed dramatically. The whole signal from $t = -\infty$ to $t = \infty$ is needed in the HT (2.17). So, time of modulation has to be *infinite*, and seemingly, real-time communication becomes impossible. Does this mean that the AS should not be applied in practical engineering?

The answer we will come to is *no*. The AS is the only cogent way for defining the APF, and any other way violates reasonable physical conditions. Moreover, the AS is widely used in practical devices invented long before the AS (Chapter 4). Like electromagnetic waves are transformed into local light rays for short wavelengths, the global AS produces local APF *asymptotically*, for slow modulation [1,2]. Conditions of this transformation are very important for practical methods based on the local idea.

3.1 Passive and Active Chirp Signals

Chirp signals with linear FM are widespread in radar, and two very different methods for their generation are often employed. In the *passive* method, the signal is produced in a filter (Figure 3.1). It is well known [3] that, using a transversal filter (in particular with a multitap delay line), we can approximate arbitrary transfer function $K(\omega) = A(\omega)e^{i\psi(\omega)}$. If the group delay is linear:

$$\tau(\omega) = -\frac{d\psi}{d\omega} = \tau_0 + c \cdot (\omega - \omega_0) \tag{3.1}$$

$K(\omega)$ reproduces the spectrum $U(\omega)$ of a chirp signal $u(t)$. Therefore, exciting the filter with a short impulse $\delta(t)$, we generate the chirp signal. Then the signal should be amplified and transformed to the desired carrier frequency. Analogous digital methods are also employed.

In contrast, the frequency generated in the *active* method is controlled with a servo-system shown in Figure 3.2. The output chirp signal $u(t)$ is delayed by a short time δt, and then the $u(t)$ and $u(t-\delta t)$ are mixed. The *constant* frequency difference δf produced in the mixer is then compared with a reference frequency f_0 in a phase detector. Its output signal dependent on the difference $\delta f - f_0$ controls the generator. Clearly, this difference is zero if the chirp signal of the frequency rate $f_0/\delta t$ is generated.

A fundamental distinction between the two methods should be mentioned. The passive method is global. The filter in Figure 3.1 knows the signal wanted. From characteristics $A(\omega)$ and $\psi(\omega)$ of the filter, one can *beforehand* determine what a signal $u(t)$ will appear at each time. On the other hand, nothing in Figure 3.2 knows the signal beforehand. Instead, the frequency difference δf is measured, and the generator is tuned during a short time δt around each instant. So, the passive method is based on a spectral (global) approach, whereas the active method uses the local frequency slowly varying in time. In spite of

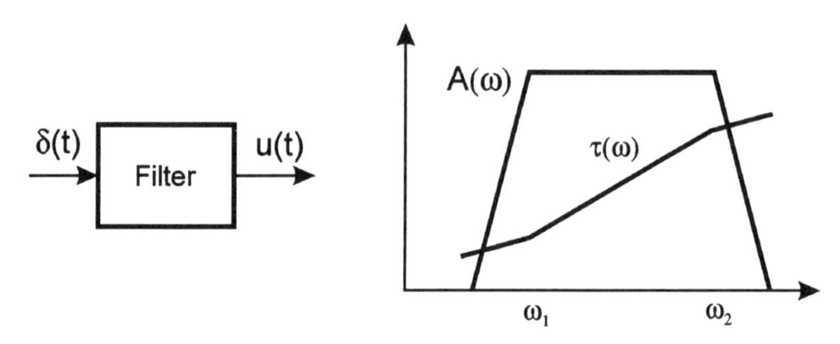

Figure 3.1 Passive method.

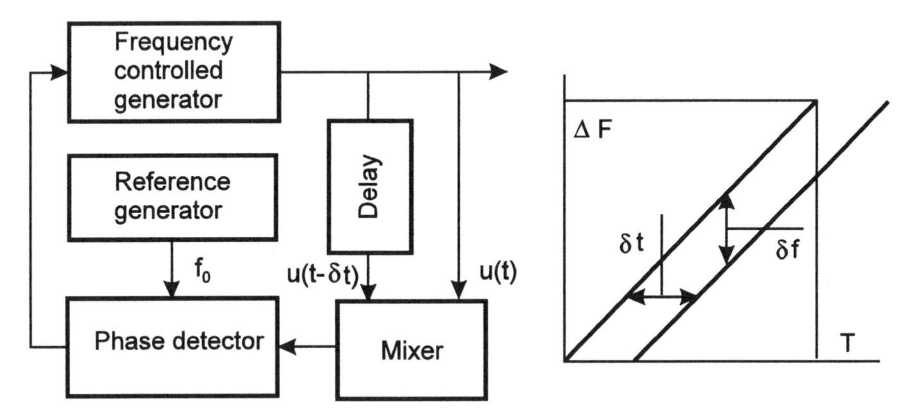

Figure 3.2 Active method.

this, the signals generated are practically identical: the same modulation appears in a frequency detector, and radar echoes are also identical. Why is it so? What conditions should be met to be so?

3.2 Stationary Phase Method and Optics

Before answering these questions, let us consider an approximate method for integration of oscillating functions as follows:

$$\int_{-\infty}^{\infty} A(x)e^{i\phi(x)}\,dx \tag{3.2}$$

Generally, adjacent half-waves of oscillations cancel each other. However, oscillations are stopped at the *stationary point* x_s where frequency is zero,

$$\phi'(x_s) = 0 \tag{3.3}$$

A neighborhood of this point makes a main contribution to the integral.

Within the neighborhood, keeping terms up to the second order of Taylor expansion for ϕ, we can set

$$\phi(x) \approx \phi_s + \phi_s'(x - x_s) + \frac{\phi_s''}{2}(x - x_s)^2 \qquad A(x) \approx A_s \tag{3.4}$$

where a subscript s denotes functions taken at the stationary point, and according to (3.3), $\phi_s' = 0$. Then we obtain the *stationary phase approximation* (SPA) for the integral (3.2):

$$\int_{-\infty}^{\infty} A(x)e^{i\phi(x)}\,dx \sim \int_{-\infty}^{\infty} A(x_s)e^{i[\phi_s+\frac{1}{2}\phi_s''(x-x_s)^2]}\,dx = \sqrt{\frac{2\pi i}{\phi_s''}}\,A_s\,e^{i\phi_s} \tag{3.5}$$

Here $\sim$ denotes *asymptotic* approximation, for fast oscillating functions. We emphasize that the integral (a global notion) depends in (3.5) on a local behavior at a single point x_s.

The size δx of the essential neighborhood surrounding this point can be estimated as follows. Let us consider the integral (3.2) within finite limits, from $x_s - \delta x$ to $x_s + \delta x$. Then, using the same approximation (3.4), we come to the Fresnel integral:

$$A_s\,e^{i\phi_s}\int_{-\delta x}^{\delta x} e^{i\frac{1}{2}\phi_s''x^2}\,dx \tag{3.6}$$

which approaches (3.5) if the δx is as follows (see Problem 3.1):

$$\delta x \sim \frac{1}{\sqrt{|\phi_s''|}} \tag{3.7}$$

In optics, for reflection from a mirror, the stationary point (3.3) is the point of reflection, and (3.7) is the size of the Fresnel zone around this point. When the Fresnel zone is on a mirror surface, (3.5) is true, and we observe the reflection pattern predicted with geometric optics. However, if the Fresnel zone happens on the edge of a mirror, we observe scattering from the edge. (See details in Problem 3.1.) If a small mirror is less than the Fresnel zone, local geometric optics is completely wrong. So, because of diffraction phenomena, a small parabolic reflector does not form a pencil of rays. For the same reason, a microscope cannot resolve objects within the Fresnel zone. Both limitations are of one nature: the neighborhood (3.7) of the stationary point must be inside the interval of integration, but not at its edge or outside.

3.3 Stationary Phase Method and FM Signals

Now we return to the passive method in Figure 3.1. Characteristics of the filter are known, and the generated real signal is as follows:

$$u(t) = \frac{1}{2\pi}\int_{-\infty}^{\infty} A(\omega)e^{i[\psi(\omega)+\omega t]}\,d\omega$$

Therefore, according to (2.12), restricting the spectrum to positive frequencies, we come to the AS:

$$w(t) = \frac{1}{\pi}\int_0^{\infty} A(\omega)e^{i[\psi(\omega)+\omega t]}\,d\omega \tag{3.8}$$

From this integral, one can find *exact* global amplitude and frequency of a signal. However, if the exponent in (3.8) is fast oscillating, the SPA is also applicable for the *approximate* calculation. Then we come to local notions.

According to (3.3) and (3.8), the stationary point ω_s is given by

$$\frac{d}{d\omega}[\psi(\omega) + \omega t] = 0 \quad \text{or} \quad -\psi'(\omega_s) = \tau(\omega_s) = t \tag{3.9}$$

and, for a given t, the signal is essentially defined by one frequency ω_s in its spectrum. This is the frequency in which the group delay $\tau(\omega_s)$ equals a given t. From (3.5), we obtain the approximate AS as follows:

$$w(t) \sim \sqrt{\frac{i}{2\pi\psi''(\omega_s)}}A(\omega_s)e^{i[\psi(\omega_s)+\omega_s t]} \tag{3.10}$$

Then, also approximately, its instantaneous frequency takes the form

$$\omega(t) = \frac{d}{dt}[\psi(\omega_s) + \omega_s t] = [\psi'(\omega_s) + t]\frac{d\omega_s}{dt} + \omega_s = \omega_s \tag{3.11}$$

where (3.9) has been taken into account.

We have obtained important relations. In general, a signal produced in a filter depends on its frequency response at $-\infty < \omega < \infty$, and for the global AS, amplitude and frequency depend on time behavior at $-\infty < t < \infty$. Nevertheless *approximately*, the signal is defined by the alone frequency ω_s, which is the spectral and instantaneous frequency for a given t. In other words, the signal carries only one frequency at each time.

Just the same characterizes the active method in Figure 3.2. The generator tuned with a servo-system of small operating time δt produces one frequency at each time, which is the spectral and instantaneous frequency as well. So, the common local idea of a slowly varying frequency does not conflict with but approximates the global AS under a certain condition.

What is that condition? For the integral (3.8), the size (3.7) is given by

$$\delta\omega \sim \frac{1}{\sqrt{|\psi_s''|}} = \frac{1}{\sqrt{\frac{d\tau_s}{d\omega}}} = \sqrt{\frac{d\omega(t)}{dt}} \tag{3.12}$$

where interrelations (3.9) and (3.11) between frequency and time domains have been taken into account again. This neighborhood must be inside the filter's band, which is the actual interval of integration in (3.8), so that $\omega_1 + \delta\omega < \omega(t) < \omega_2 - \delta\omega$. Here, ω_1 and ω_2 are the bounds shown in Figure 3.1.

Further, the frequency interval $\delta\omega$ corresponds to the time interval

$$\delta t = \delta\omega \cdot \frac{dt}{d\omega} \sim \sqrt{\frac{d\omega}{dt}} \cdot \frac{dt}{d\omega} = \sqrt{\frac{dt}{d\omega}} = \sqrt{\frac{T}{\Delta\omega}} \tag{3.13}$$

Here, T is duration and $\Delta\omega$ is frequency deviation of a signal. The interval (3.13) must be inside signal duration, so that $\delta t < t < T - \delta t$.

In Figure 3.2, the δt is delay, and the servo-system cannot operate properly if δt happens at the edge of a signal. In fact, either $u(t)$ or $u(t - \delta t)$ disappears near the edge, and comparison in the mixer cannot be done. Like for geometric optics, then, the edge effect plays a fundamental role, and the local frequency notion is right for *long* signals with respect to δt. We come to a final condition:

$$\delta t \approx \sqrt{\frac{T}{\Delta\omega}} \ll T \quad \text{or} \quad \Delta\omega \cdot T \gg 1 \tag{3.14}$$

The product $\Delta\omega \cdot T$, named the *base of a signal* (for chirp signals) or the *FM index* (for sinusoidal frequency modulation of a period T), must be large. Therefore, modulation must be slow and wideband. Under this condition, the frequency produced with the local active device is the same as for the AS in the global passive device. Besides, the instantaneous and spectral frequencies are the same at each time, and the common idea of a slowly varying frequency is right.

In signal theory, the SPA was first used as an approximate method for analysis and synthesis of FM signals [4,5]. Its physical significance is more fundamental: the SPA justifies local comprehension of signals and clarifies the condition of its applicability.

3.4 Paradoxes of Instantaneous Frequency

In spite of both being measured in hertz, spectral and instantaneous frequencies are different concepts. If we confuse them for fast or narrowband modulation, paradoxes and misunderstanding may occur.

Paradox of Two Sinusoids A sum of two sinusoids $u(t) = a_1 \cos\omega_1 t + a_2 \cos\omega_2 t$ is the simplest multiharmonic signal. Its AS amplitude and instantaneous frequency are given by

$$a(t) = \sqrt{a_1^2 + a_2^2 + 2a_1 a_2 \cos(\omega_2 - \omega_1)t}$$

$$\omega(t) = \frac{\omega_1 a_1^2 + \omega_2 a_2^2 + (\omega_1 + \omega_2)a_1 a_2 \cos(\omega_2 - \omega_1)t}{a_1^2 + a_2^2 + 2a_1 a_2 \cos(\omega_2 - \omega_1)t}$$

whereas spectral frequencies are ω_1 and ω_2. As shown in Figure 3.3, the instantaneous frequency differs from spectral ones and even becomes negative. Why is this so?

The modulating frequency $\omega_2 - \omega_1$ is of the same order as deviation, and $\Delta\omega T \sim 1$. So, the condition (3.14) is violated, and the instantaneous frequency may differ from spectral frequencies.

Paradox of Narrowband Frequency Modulation [6,7] As early as the 1920s, radio bands became tight for many radio channels, and in 1929 Robinson took out the U.S. Patent for narrowband FM instead of AM. His reasoning was as follows. For 3-kHz speech modulation, each AM channel takes 6 kHz. Using FM, however, one can narrow the frequency deviation down to 1 kHz or less. Then each channel will take a 1-kHz narrowband.

This reasoning is correct if the instantaneous and spectral frequencies are the same. For narrowband modulation, however, the FM index is small, and condition (3.14) is violated. Therefore, local interpretation is irrelevant, and the spectral band is wider than frequency deviation.

Even earlier, the attempts were made to produce frequency modulation with a capacitive microphone connected with a resonant circuit in a generator.

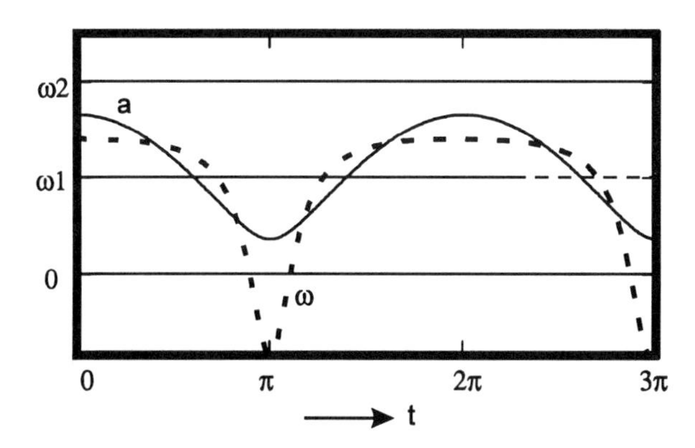

Figure 3.3 Amplitude (solid line) and frequency (dotted line) of the biharmonic signal $u(t) = \cos t + 0.65 \cos 2t$.

The results were pure, however, and frequency detectors could not reproduce modulation. The problem was the same. For long-wave generators of that time, a varying capacity of a microphone was small, and the FM index was insufficient for local handling typical for frequency detectors. We will show in Chapter 4 that the global AS was employed in the first successful FM system.

Paradox of Rectangular Envelope The AS amplitude (envelope) of a rectangular radio impulse is shown in Figure 3.4. It starts before and finishes after the real signal. Puzzling questions arise: What does the amplitude envelop when no signal exists? Why does the amplitude outstrip the signal though the output of a detector does not? Why does the envelope differ from a rectangle?

Just as light leaks behind an aperture, the global amplitude leaks behind a rectangle. In practical detectors, output signals are also longer than input ones. The output does not outstrip the input because of additional delay in a detector, however.

3.5 Alternative Methods

So, local amplitude and frequency are approximate notions, whereas global AS defines exact APF. This, however, is improper for many people, and in the past 60 years, alternative methods have been proposed. All the methods are *local*, and the APFs depend on signal behavior at a given t. We discuss the proposed APFs and show their imperfections. We also discuss fast oscillating quadratures typical for alternative methods.

Preliminarily, however, we formulate reasonable physical conditions for the APF and show that only the AS meets them [8]. In the consideration, the following properties of operators will be employed.

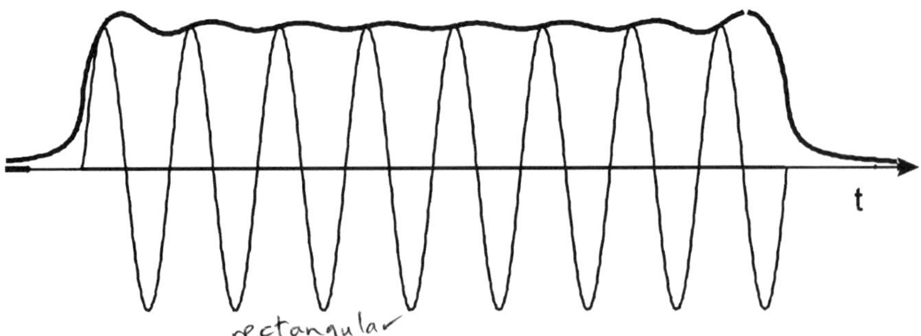

Figure 3.4 Paradox of envelope.

Operators and Their Derivatives As applied to signals, operators describe transformations of input signals into output ones. We write $u_{out}(t) = \mathcal{A}[u_{in}(t)]$, and, for example, the operator $\mathcal{A}$ may describe a filter $K(\omega)$ or a quadratic detector. Then the operators are given by

$$\mathcal{A}(u) = \mathcal{F}^{-1}[K\mathcal{F}(u)] \quad \text{or} \quad \mathcal{A}(u) = |\mathcal{I}(u) + i\mathcal{H}(u)|^2$$

In the first relation, $\mathcal{F}^{-1}$ and $\mathcal{F}$ are inverse and direct Fourier transforms (and are also operators). In the second relation, we conclude that the detector extracts the envelope of the AS (see Chapter 4), so that $\mathcal{I}$ and $\mathcal{H}$ are the unit operator and the HT, respectively (see Problem 2.11). Using the input spectrum $U(\omega) = \mathcal{F}(u)$, these operators can be written explicitly:

$$\mathcal{A}(u) = \frac{1}{2\pi}\int_{-\infty}^{\infty} K(\omega)U(\omega)e^{i\omega t}\,d\omega \quad \text{or} \quad \mathcal{A}(u) = \left|\frac{1}{\pi}\int_{0}^{\infty} U(\omega)e^{i\omega t}\,d\omega\right|^2$$

Properties of operators are often similar to those of functions. So, the operator is *linear* if $\mathcal{A}(c_1u_1 + c_2u_2) = c_1\mathcal{A}(u_1) + c_2\mathcal{A}(u_2)$ for any numbers c_1, c_2 and any functions u_1, u_2. Clearly, the first operator above is linear, but the second operator is nonlinear. Further, the operator is *homogeneous* if $\mathcal{A}(cu) = c\mathcal{A}(u)$ for a constant c. Also, the operator is *continuous* if a small variation of u results in a small variation of $\mathcal{A}(u)$ so that

$$\mathcal{A}(u + \delta u) \rightarrow \mathcal{A}(u) \quad \text{for } \|\delta u\| \rightarrow 0$$

Here $\|\delta u\|$ denotes the norm. For a given signal, the norm is a constant often associated with energy of the signal:

$$\|\delta u\|^2 = \int_{-\infty}^{\infty} \delta u^2(t)\,dt$$

So, instead of the condition $\delta u \rightarrow 0$ for functions, we use the condition $\|\delta u\| \rightarrow 0$ for operators, which means that the energy of $\delta u(t)$ runs to zero.

Like for functions, the *differentiable* operator meets the condition

$$\mathcal{A}(u + \delta u) = \mathcal{A}(u) + \mathcal{A}'(u)\delta u + \varepsilon(u, \delta u)$$

Here, the derivative $\mathcal{A}'(u)$ is a linear operator acting on δu but dependent on u as well. The operator $\varepsilon(u, \delta u)$ is also acting on δu but is of a higher order. This means that energy of $\varepsilon(t)$ approaches zero faster than that of $\delta u(t)$. This condition is formally written in (3.20) below.

3.5.1 Physical Conditions for the APF

According to Section 2.1, the APF are given by

$$a(t) = \sqrt{u^2(t) + v^2(t)} = |w(t)| \tag{3.15}$$

$$\phi(t) = \arctan\left\{\frac{v(t)}{u(t)}\right\} = \operatorname{Im}\{\ln w(t)\} \tag{3.16}$$

$$\omega(t) = \frac{d\phi}{dt} = \frac{v'(t)u(t) - u'(t)v(t)}{u^2(t) + v^2(t)} = \operatorname{Im}\left\{\frac{w'(t)}{w(t)}\right\} \tag{3.17}$$

where $v(t) = \mathcal{H}[u(t)]$ is a conjugated signal produced with an operator $\mathcal{H}$. If $\mathcal{H}$ is the HT (2.17), we come to the AS.

We now apply reasonable physical conditions to the APF, which the operator $\mathcal{H}$ must satisfy, and show that the HT is the only operator that satisfies them.

Condition A: Amplitude Continuity Let a small variation $\delta u(t)$ be added to a signal $u(t)$. Clearly, the associated amplitude variation $\delta a(t)$ must also be small, and since the transformation (3.15) is continuous, the same applies to $\delta v(t)$. Thus we conclude that the operator $\mathcal{H}$ must be continuous:

$$\mathcal{H}[u + \delta u] \to \mathcal{H}[u] \quad \text{for } \|\delta u\| \to 0 \tag{3.18}$$

In the space of continuous operators, the set of differentiable operators is *dense* [22]. In other words, allowing an arbitrary small change for any signal, we can replace a continuous operator $\mathcal{H}$ by a differentiable operator, such that

$$\mathcal{H}[u + \delta u] = \mathcal{H}[u] + \mathcal{H}'[u]\delta u + \varepsilon[u, \delta u] \tag{3.19}$$

By definition [22], the derivative $\mathcal{H}'(u)$ exists, and $\mathcal{H}(u)$ is differentiable if there exists such an operator $\varepsilon(u, \delta u)$ in (3.19) that approaches zero faster than δu:

$$\frac{\|\varepsilon[u, \delta u]\|}{\|\delta u\|} \to 0 \quad \text{for } \|\delta u\| \to 0 \tag{3.20}$$

The condition of differentiability looks more sophisticated than that of continuity. Note, however, that for the derivative $v'(t)$, we have

$$\begin{aligned} v'(t) &= \lim_{\delta t \to 0} \frac{v(t + \delta t) - v(t)}{\delta t} = \lim_{\delta t \to 0} \frac{\mathcal{H}[u(t) + \delta u(t)] - \mathcal{H}[u(t)]}{\delta t} \\ &= \lim_{\delta u \to 0} \frac{\mathcal{H}[u + \delta u] - \mathcal{H}[u]}{\delta u} \lim_{\delta t \to 0} \frac{\delta u}{\delta t} = \mathcal{H}'(u)u'(t) \end{aligned}$$

where (3.19) and (3.20) are taken into account, and we take such $u(t)$ and $\delta u(t)$ that $\mathcal{H}[u(t)] = v(t)$ and $\mathcal{H}[u(t) + \delta u(t)] = v(t + \delta t)$. Therefore, existence of the derivative $\mathcal{H}'[u]$ is directly required in (3.17) for the frequency to exist.

Condition B: Phase Independence of Scaling and Homogeneity Let a signal $u(t)$ be replaced by $cu(t)$ for a constant $c > 0$. Then phase and frequency must remain the same, and by virtue of (3.16), $\mathcal{H}$ is hence a homogeneous operator:

$$\frac{\mathcal{H}[cu]}{cu} = \frac{\mathcal{H}[u]}{u}$$

that is,

$$\mathcal{H}[cu] = c\mathcal{H}[u] \tag{3.21}$$

Condition C: Harmonic Correspondence The constant amplitude and frequency of a simple sinusoid must retain their values. Therefore, according to (3.15) and (3.16), for *any constant* $a > 0$, $\omega > 0$, and Φ, we must have

$$\mathcal{H}[a\cos(\omega t + \Phi)] = a\sin(\omega t + \Phi) \tag{3.22}$$

Condition C is the same as property 2 AS. In Section 2.2, we have seen that the only *linear* (additive) operator that satisfies this condition is the HT. In Problem 3.2, we also show that the operator must be linear to meet conditions A and B.

Therefore, only the HT meets all three conditions, and the AS takes a special position among APF definitions. Any other way of defining the amplitude and phase means that at least one condition is violated. For example, the amplitude may be discontinuous or not constant for a simple sinusoid.

3.5.2 Fast Oscillating Quadratures

In Section 2.3, we considered slow *band-limited quadratures* (BLQ) of a signal. The quadratures $x(t)$ and $y(t)$ are related to real signals as follows (we set $\omega_0 = 1$):

$$u(t) = a(t)\cos[t + \Phi(t)] = x(t)\cos t - y(t)\sin t \tag{3.23}$$

and they result in the amplitude and phase according to (2.19). We have also shown that, because of Bedrosian's theorem, the AS gives the same APF.

However, even for a narrowband signal, the functions $x(t)$ and $y(t)$ in (3.23) need not be band-limited and can involve oscillations of double the carrier frequency. If we let $x(t)$ and $y(t)$ be the BLQ, then the new quadratures $\xi(t)$ and $\eta(t)$ are given by

$$\begin{aligned}\xi(t) &= x(t) + f(t) - f(t)\cos 2t + g(t)\sin 2t \\ \eta(t) &= y(t) + g(t) + f(t)\sin 2t + g(t)\cos 2t\end{aligned} \tag{3.24}$$

where $f(t)$ and $g(t)$ are arbitrary. If we substitute ξ and η into (3.23) instead of x and y, fast oscillations cancel each other, and the initial signal is restored due to the trigonometric identity

$$\begin{aligned}&(x + f - f\cos 2t + g\sin 2t)\cos t - (y + g + f\sin 2t + g\cos 2t)\sin t \\ &\quad = x\cos t - y\sin t\end{aligned} \tag{3.25}$$

This shows that quadratures are ambiguous as well as the amplitude and phase.

Let us consider an example. The BLQ of a simple sinusoid $u(t) = \cos t$ are $x = 1$ and $y = 0$, and its amplitude and frequency are $a = 1$ and $\omega = 1$. Figure 3.5 shows, however, the APF of the same sinusoid associated with oscillating quadratures (3.24) for $f = g = 0.25$. Fast varying amplitude and frequency obtained differ significantly from $a = 1$ and $\omega = 1$, and condition C is violated. However, by virtue of (3.25), the product $a(t)\cos\phi(t)$ *exactly restores* a sinusoidal signal.

Thus, contrary to our intuition, fast oscillations of double the carrier frequency are formally admissible in the sense that a signal can be restored from the oscillating APF. Alternative methods are of this kind and display fast distortions. We also emphasize that additional averaging (filtering) for removing undesirable oscillations is ineffective. Indeed, the averaged amplitude in Figure 3.5 also

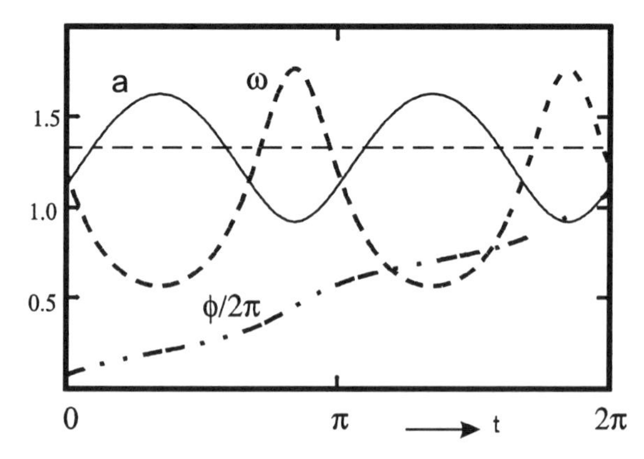

Figure 3.5 Fast varying APF for a simple sinusoid $u(t) = \cos t$ (the dash-dot straight line shows the averaged amplitude).

differs from $a = 1$. This comment is important for the following discussion of Mandelstam's method.

3.5.3 Alternative APF Definitions

Alternative APF methods were suggested before and after 1946 when the AS was introduced. All implicitly involve fast varying quadratures and violate the conditions of Section 3.5.1.

Mandelstam's Method This method was originated in 1934 and is widely used in nonlinear oscillation theory [9–15]. Using a representation on the phase plane $(u, -u')$, Mandelstam defined the amplitude and phase as polar coordinates on it; that is,

$$\begin{aligned} u &= a\cos(t+\Phi) = \xi\cos t - \eta\sin t \\ -u' &= a\sin(t+\Phi) = \xi\sin t + \eta\cos t \end{aligned} \tag{3.26}$$

The conjugated signal is $\mathcal{H}[u(t)] = -u'(t)$, and this linear transformation meets conditions A and B. For a sinusoid $u = \cos\omega t$, however, we have $-u' = \omega\sin\omega t$, and condition C is met for $\omega = 1$ only. Sinusoids of other frequencies are not transformed according to (3.22). Mandelstam's method is a special case of (3.24). Indeed, using $u(t)$ and its derivative from (3.23), we can solve the system (3.26) and show that ξ and η are given by (3.24) with arbitrary functions specified as $f(t) = y'(t)/2,\ \ g(t) = -x'(t)/2$.

For any signal, except a sinusoid of unit frequency, Mandelstam's APF oscillate. Carrying out this procedure typically involves averaging. This deletes fast variations, but distortions nonetheless remain as illustrated in Figure 3.5. This limits accuracy to the first or zeroth order. We will show in Chapters 6 and 7 that the AS provides more accurate results for typical oscillation problems. Figure 3.6 shows Mandelstam's amplitude and frequency for a chirp signal $u(t) = \cos(t+\mu t^2)$. Fast distortions, especially in frequency, increase when $\omega(t)$ moves away from $\omega = 1$.

Shekel's Method This method [16] looks very reasonable. Around a given point, Shekel approximates a signal $u(t)$ with a simple sinusoid $a\cos(\omega t+\Phi)$. Equating the signal and its first and second derivatives to the corresponding values of the sinusoid:

$$\begin{cases} u(t) &= a\cos(\omega t+\Phi) \\ u'(t) &= -a\omega\sin(\omega t+\Phi) \\ u''(t) &= -a\omega^2\cos(\omega t+\Phi) \end{cases}$$

and solving for amplitude and frequency, Shekel obtains

$$a(t) = \sqrt{u(t)^2 - \frac{u'(t)^2 u(t)}{u''(t)}} \qquad \omega(t) = \sqrt{-\frac{u''(t)}{u(t)}} \tag{3.27}$$

In contrast to Mandelstam's method, the amplitude and frequency are correct for *any sinusoid*, and condition C is met. Since frequency is independent of scaling, condition B is also met. Condition A is violated, however, and in general the amplitude is infinite for $u''(t) = 0$.

For the same chirp signal used in Figure 3.6, Shekel's amplitude and frequency are shown in Figure 3.7. Distortions are significant and unusual because, due to modulation, formulas (3.27) contain tangent under the roots. Twice a period, the tangent takes infinite values, and at small parts of a period, the amplitude and frequency are imaginary. (See Problem 3.3 for another example.) This shows that Shekel's method is inapplicable for excepting strict sinusoids

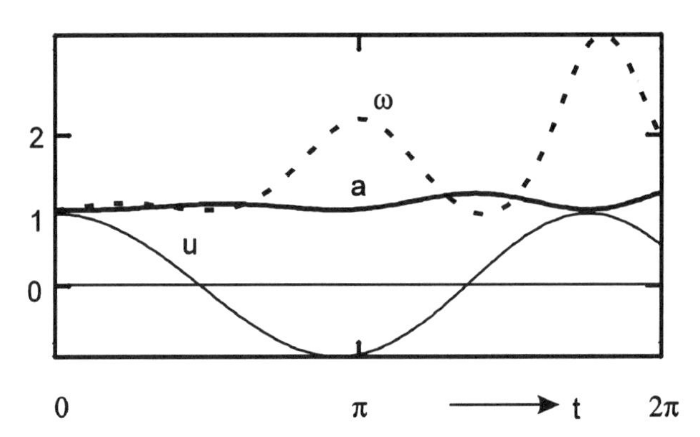

Figure 3.6 Chirp signal $u(t) = \cos(t+\mu t^2)$ for $\mu = 0.025$ and $0 < t < 2\pi$ and its Mandelstam's amplitude and frequency.

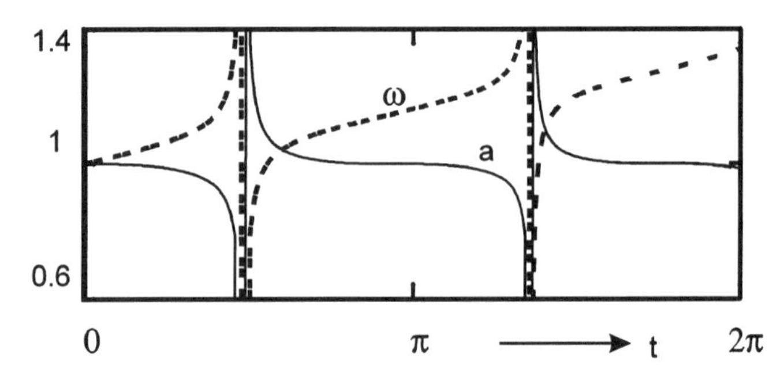

Figure 3.7 Shekel's amplitude and frequency for the same chirp signal.

and that reasonable APF cannot be obtained by local approximation at a given instant. We consider the approximation over a finite interval in Problem 3.6. It is also unsatisfactory.

Teager-Kaiser Energy Algorithm (TKA) This method has recently been introduced for speech analysis as a simple definition of amplitude and frequency [17–19]. For narrowband noise-free signals, the method is easy and effective. It is especially effective for sampled signals.

For a sinusoid $u(t) = a\cos(\omega t + \Phi)$ and its derivatives, the following relations are valid:

$$\Psi(u) = u'(t)^2 - u(t)u''(t) = a^2\omega^2 \tag{3.28}$$

$$\Psi(u') = u''(t)^2 - u'(t)u'''(t) = a^2\omega^4 \tag{3.29}$$

and the TKA admits them generally. Then the amplitude and frequency of the arbitrary signal are as follows:

$$a(t) = \frac{\Psi(u)}{\sqrt{\Psi(u')}} \qquad \omega(t) = \sqrt{\frac{\Psi(u')}{\Psi(u)}} \tag{3.30}$$

As in Shekel's method, conditions B and C are satisfied but condition A is violated. Nevertheless, the results are much better for most signals. For the same chirp signal, the amplitude and frequency are shown in Figure 3.8. Fast variations (again of the double carrier frequency) are small, and the amplitude and frequency are suitable.

However, examples that give poor results can be constructed. The FM signal $u(t) = \sin(t + t^3/6)$ of a quadratic frequency $\omega(t) = 1 + t^2/2$ has the power series $u(t) = t - \frac{3}{40}t^5 + \cdots$. Therefore, its derivatives are $u'(0) = 1$, $u''(0) = 0$,

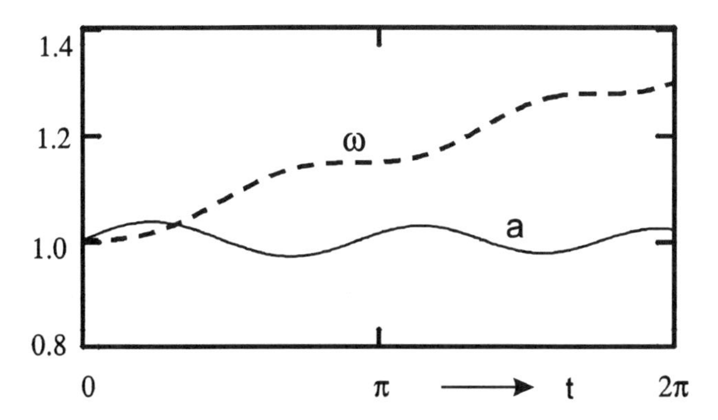

Figure 3.8 Teager-Kaiser's amplitude and frequency for the same signal.

$u'''(0) = 0$, and, according to (3.28) and (3.29), $\Psi(u) = 1$ and $\Psi(u') = 0$ for $t = 0$. Then the amplitude (3.30) is infinite and the frequency is zero. This FM signal and its TKA and AS amplitude and frequency are shown in Figure 3.9. The AS meets our intuition, whereas the TKA gives wrong results for $t = 0$. Problem 3.4 gives similar examples.

Summarizing, we conclude that the alternative methods violate important physical conditions and display fast distortions in the APF. The TKA is the best among alternative methods. It shows acceptable small distortions for narrowband signals but produces singularities for special wideband signals. Local alternative methods extract the APF from a signal taken at a given time, whereas for the global AS, a finite (or even infinite) interval is necessary. Nevertheless, removing fast distortions and singularities, the AS provides reasonable APFs, which are as slow as possible.

3.6 Semilocal Representation

Fourier spectrum is a global notion: the whole spectrum at $-\infty < \omega < \infty$ is needed for signal reconstruction at each t:

$$w(t) = \frac{1}{2\pi} \int_{-\infty}^{\infty} W(\omega) e^{i\omega t}\, d\omega \tag{3.31}$$

and the whole signal at $-\infty < t < \infty$ defines the spectrum. However, the SPA shows that a main contribution into a signal is made by a neighborhood of the instantaneous frequency $\omega(t)$, so that

$$w(t) \approx \frac{1}{2\pi} \int_{\omega(t)-B}^{\omega(t)+B} W(\omega) e^{i\omega t}\, d\omega \qquad B \approx \sqrt{\left| \frac{d\omega(t)}{dt} \right|} \tag{3.32}$$

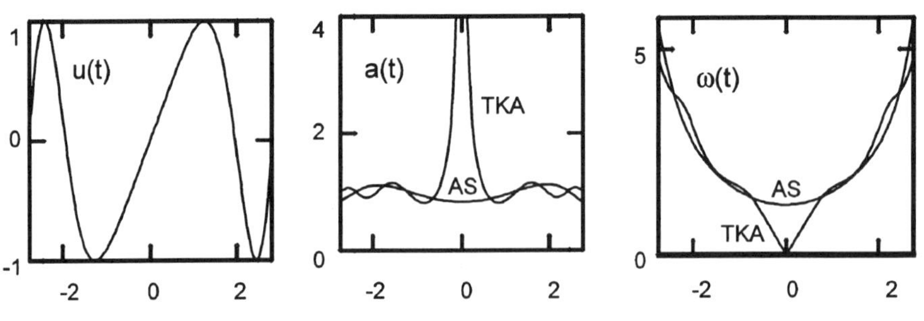

Figure 3.9 The AS and TKA amplitude and frequency for the wideband FM signal $u(t) = \sin(t + t^3/6)$.

The SPA is an approximate method. However, we now develop *exact reconstruction* of a signal from its *semilocal* spectrum around the instantaneous frequency [20]. First we consider arbitrary complex signals, but then we will show that the AS is preferred for this representation.

Given a complex signal $w(t) = a(t)e^{i\phi(t)}$, we introduce its *complex phase* $\theta(\mathrm{t})$ as follows:

$$w(t) = a(t)e^{i\phi(t)} = e^{i\theta(t)} \qquad \theta(t) = -i\ln[w(t)] = \phi(t) - i\ln[a(t)] \tag{3.33}$$

This complex phase includes not only the real phase $\phi(t)$, but also the amplitude $a(t)$. Then, like in (3.4), we *approximate* the complex phase and the signal around a given t:

$$\tilde{\theta}(t+\tau) \approx \theta_t + \theta'_t \cdot \tau + \frac{\theta''_t}{2} \cdot \tau^2 \tag{3.34}$$

$$\tilde{w}(t+\tau) \approx e^{i\left[\theta_t + \theta'_t \cdot \tau + \frac{\theta''_t}{2} \cdot \tau^2\right]} = w(t)e^{i\left[\theta'_t \cdot \tau + \frac{\theta''_t}{2} \cdot \tau^2\right]} \tag{3.35}$$

A subscript t denotes functions taken at t, and $e^{i\theta_t} = w(t)$.

Further, differentiating the $\theta(t)$ in (3.33), we introduce the *complex* instantaneous frequency θ'_t and the frequency rate θ''_t as follows (both include FM and AM):

$$\theta'_t = -i\frac{w'(t)}{w(t)} = \omega(t) - i\frac{a'(t)}{a(t)} \tag{3.36}$$

$$\theta''_t = -i\left[\frac{w''(t)}{w(t)} - \frac{w'(t)^2}{w(t)^2}\right] = \omega'(t) - i\left[\frac{a''(t)}{a(t)} - \frac{a'(t)^2}{a(t)^2}\right] \tag{3.37}$$

Finally, we define the *semilocal spectrum* as follows:

$$W_t(\omega) = K_t(\omega - \theta'_t)\int_{-\infty}^{\infty} G_t(\tau)\tilde{w}(t+\tau)e^{-i\omega(t+\tau)}\,d\tau \tag{3.38}$$

Here, $\tilde{w}$ is the approximated signal (3.35), while $K_t(\omega)$ and $G_t(\tau)$ are the windows concentrated around $\omega = 0$ and $\tau = 0$. We take the windows as follows (see also Problem 3.7):

$$G_t(\tau) = \sqrt{1+\gamma_t}\,e^{-\tau^2/T_t^2} \qquad K_t(\omega) = e^{-\omega^2/B_t^2} \tag{3.39}$$

where the parameters T_t, B_t, and γ_t are given by

$$B_t = \beta_t\sqrt{2\theta_t''} = b\sqrt{2|\theta_t''|} \qquad T_t = \frac{2\alpha}{B_t} \qquad \gamma_t = \frac{1}{\alpha^2} + \frac{i}{\beta_t^2} \tag{3.40}$$

Here, α and b are positive numbers ($\alpha < b$ for convergence), and $|\beta_t| = b$, but the phase of β_t depends on a signal at each t. The windows define a semilocal representation in the sense that $G_t(\tau)$ restricts a signal around a given t within the duration T_t, whereas $K_t(\omega - \theta_t')$ restricts its spectrum around the instantaneous frequency $\omega(t)$ within the band B_t.

For the functions (3.35) and (3.39), the semilocal spectrum (3.38) can be found explicitly:

$$\begin{aligned} W_t(\omega) &= \sqrt{1+\gamma_t}\, e^{-(\omega-\theta_t')^2/B_t^2} w(t) e^{-i\omega t} \\ &\times \int_{-\infty}^{\infty} \exp\left[-\frac{\tau^2}{T_t^2} + i\left(\theta_t'\tau + \theta_t''\frac{\tau^2}{2} - \omega\tau\right)\right] d\tau \\ &= \frac{2\sqrt{\pi}}{B_t}\sqrt{\frac{1+\gamma_t}{\gamma_t}}\, w(t) e^{-i\omega t} \exp\left\{-\frac{1+\gamma_t}{\gamma_t}\frac{(\omega-\theta_t')^2}{B_t^2}\right\} \end{aligned} \tag{3.41}$$

Concentrated around the complex instantaneous frequency θ_t', this spectrum occupies the band $B_t \sim \sqrt{|\theta_t''|}$ dependent on both FM and AM at each time.

The amazing fact is that, in spite of the approximation (3.34), the spectrum (3.41) provides *exact reconstruction* of a signal. Indeed, applying Fourier transform to (3.41), we find

$$\begin{aligned} \frac{1}{2\pi}\int_{-\infty}^{\infty} W_t(\omega) e^{i\omega t}\, d\omega &= \frac{w(t)}{\sqrt{\pi}\, B_t}\sqrt{\frac{1+\gamma_t}{\gamma_t}} \int_{-\infty}^{\infty} \exp\left\{-\frac{1+\gamma_t}{\gamma_t}\frac{\omega^2}{B_t^2}\right\} d\omega \\ &= w(t) \end{aligned} \tag{3.42}$$

We come to *tracking reception* of signals. The receiver of the frequency response $K_t(\omega)$ follows signal frequency at each instant, and its band B_t depends on the instantaneous FM and AM rates. Besides, the receiver processes a part of the signal within the window $G_t(\tau)$ around each t, and duration of this part depends on the instantaneous band. Then exact reconstruction is achieved. So two methods of reception are feasible. The common receiver takes in the whole spectrum (3.31), but the tracking receiver extracts a part of the spectrum around the instantaneous frequency. Theoretically, both methods are exact.

Relation to the AS We did not specify the complex signal, and $w(t)$ may differ from the AS. However, the AS is preferred for two reasons.

The semilocal spectrum follows the instantaneous frequency. For a chirp signal, this spectrum is a ridge along the linear frequency $\omega(t) = \omega_0 + \mu t$. If we use the alternative $w(t)$ different from the AS, however, fast oscillations appear with double the carrier frequency as shown in Figures 3.6 through 3.8. Then the ridge oscillates, and the band B_t is wider than that for the AS. This impairs reception in a noisy background.

Another reason relates to the complex phase (3.33) and its derivatives (3.36) and (3.37), included in the semilocal spectrum. For $a = 0$, $\ln[a]$ and its derivatives are singular so that the band B_t becomes infinite. For the AS, however, excepting very special examples, $a(t) \neq 0$ at any finite t, and the singularity disappears even for time-limited signals of finite duration. So, the paradoxical amplitude in Figure 3.4 becomes helpful.

Time-Frequency Distribution For a given t, the $W_t(\omega)$ is a "representative part" of a spectrum in the sense that just this part is necessary for signal reconstruction. Therefore,

$$P(t, \omega) = |W_t(\omega)|^2 \tag{3.43}$$

is a reasonable *time-frequency distribution* (TFD) that displays the energy density in time and frequency.

Classical Fourier spectrum is global in the sense that we define $W(\omega)$ from the whole signal at $-\infty < t < \infty$. On the other hand, using the TFD we try to define the density at a given frequency and at a given time, which disagrees with Fourier spectrum. Many TFDs have been proposed for combining spectral and time patterns [21], and (3.43) is one of them. Our approach is advantageous in the sense of exact reconstruction. It generalizes the periodogram and the short-time Fourier transform.

3.7 Supplementary Problems

Problem 3.1

Consider reflection from a mirror A–B as illustrated in Figure 3.10. Show that the point of reflection is the stationary point for the integral of type (3.2). Also, estimate the Fresnel zone size that makes a main contribution to a reflected field. How does scattering appear when the Fresnel zone happens at the edge of a mirror?

Solution We set $x = 0$ at the point of reflection where the angles of falling and reflection are equal. Then distances from a point x on a mirror to the primary

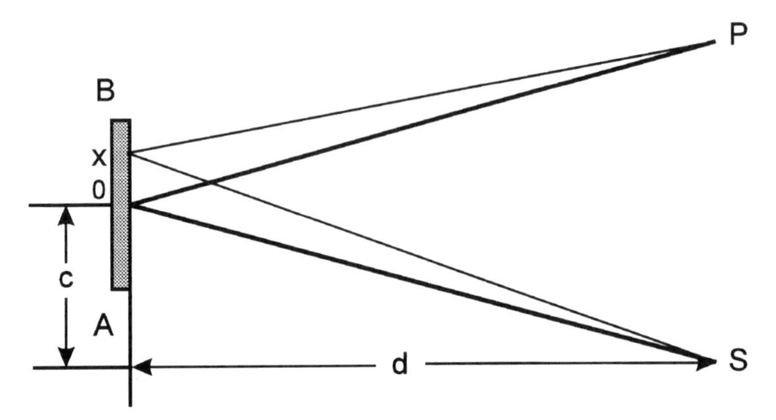

Figure 3.10 Reflection.

source S and to the point of observation P are as follows (see Figure 3.10):

$$L_1 = \sqrt{d^2 + (c+x)^2} \approx d + \frac{(c+x)^2}{2d}$$

$$L_2 = \sqrt{d^2 + (c-x)^2} \approx d + \frac{(c-x)^2}{2d}$$

Therefore, the path length from S to P through a point x is

$$L(x) = L_1 + L_2 = 2d + \frac{c^2}{d} + \frac{x^2}{d} \approx 2\sqrt{d^2 + c^2} + \frac{x^2}{d} = 2L_0 + \frac{x^2}{d}$$

where $c \ll d$ and $x \ll d$ are assumed. The minimal length $2L_0$ is for the point of reflection $x = 0$.

For a wavelength λ, the phase increment for each x is $2\pi\, L(x)/\lambda$, and each point on the mirror reflects a part of the field. Therefore, we find the reflected field as the integral over the mirror:

$$F = \int_A^B \exp\left\{ i\frac{2\pi}{\lambda}\left(2L_0 + \frac{x^2}{d}\right)\right\} dx \qquad (3.44)$$

Its stationary point (3.3) is the point of reflection $x = 0$. If this point is on a mirror surface, like in Figure 3.10, we can consider a small part of the mirror around $x = 0$, from $-\delta x$ to δx. Then we come to the Fresnel integral (3.6), and Figure 3.11 shows the amplitude $|F|$ of the reflected field as a function of δx. Clearly, the size $\delta x \sim \sqrt{\lambda d}$ produces a main part of reflection. This size defines the Fresnel zone, and only this small part of a mirror is significant for

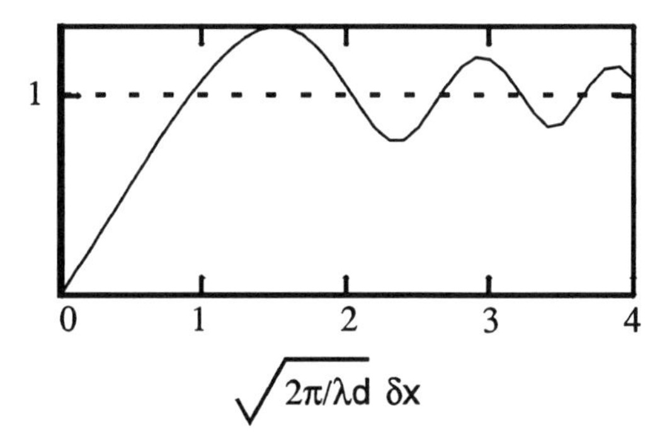

Figure 3.11 Fresnel integral.

reflection. When coming to the limit $\lambda \to 0$, we have $\delta x \to 0$. Then, according to geometric optics, light is propagating along the shortest ray L_0 (Fermat principle) and reflecting from a single point $x = 0$.

If the point $x = 0$ is outside the mirror but not far from it, a part of the Fresnel zone remaining on the mirror makes the field scattered from the edge. Scattered light cannot be explained within geometric optics but is observable as well as the other diffraction phenomena.

Problem 3.2

Show that the operator $\mathcal{H}$ in Section 3.5.1 must be linear to meet conditions A and B. Show also that continuous amplitude $a(t)$ may not be differentiable though the operator $\mathcal{H}$ is.

Solution A signal u can be written as $u = \|u\| \cdot \tilde{u}$, where $\tilde{u}$ is a point on the unit hypersphere and the norm $\|u\|$ (energy of a signal) is a constant. Further, because of homogeneity, we have that $\mathcal{H}(u) = \|u\|\mathcal{H}(\tilde{u})$, and for $\|u\| \to 0$, because of continuity, we have $\mathcal{H}(0) = 0$. Therefore, around the point 0, we rewrite (3.19) and (3.20) as follows:

$$\mathcal{H}(u) = \mathcal{H}'(0)u + \varepsilon(u) \tag{3.45}$$

where

$$\frac{\|\varepsilon(u)\|}{\|u\|} \to 0 \quad \text{for } \|u\| \to 0 \tag{3.46}$$

Here we have replaced δu with u, and therefore, the operators $\mathcal{H}'(0)$ and ε are acting on u. To prove the desired result, we have to show that $\varepsilon(u) = 0$; then (3.45) shows that $\mathcal{H}(u)$ is a linear operator.

For $\varepsilon(u) \neq 0$, we come to a contradiction in (3.45). A homogeneous operator $\mathcal{H}(u)$ and a linear operator $\mathcal{H}'(0)u$ are both of the first order with respect to $\|u\|$, but $\varepsilon(u)$ is of a higher order. Therefore, equality holds for $\varepsilon(u) = 0$ only. Formally, we can show this as follows.

According to (3.45), operator $\varepsilon(u)$, being a difference of a homogeneous $\mathcal{H}(u)$ and a linear $\mathcal{H}'(0)u$, is homogeneous itself, so that

$$\varepsilon(u) = \varepsilon(\|u\|\tilde{u}) = \|u\|\varepsilon(\tilde{u})$$

Substituting this into (3.46), we see that $\varepsilon(\tilde{u}) = 0$ because $\tilde{u}$ is independent of $\|u\|$. Hence, $\varepsilon(u) = 0$, which completes the proof.

Our reasoning was as follows. Continuous amplitude $a(t)$ results in a continuous operator $\mathcal{H}$. Then, a continuous operator was replaced by a differentiable one because of denseness, and finally, we showed that the differentiable and homogeneous operator must be linear. The amplitude itself may not be differentiable, however. It is clear from the example $u(t) = \cos\omega t \cdot \cos\omega_0 t$ considered in Section 2.3. The linear (differentiable) HT results in the AS $w(t) = \cos\omega t e^{i\omega_0 t}$, and its amplitude $a(t) = |\cos\omega t|$ is continuous but not differentiable at the points where $a(t) = 0$. Also, for these points, the phase and frequency are discontinuous, and continuity has been assumed for the amplitude but not phase or frequency.

Problem 3.3

In Section 3.5.3, we have considered distortions of Shekel's frequency for the FM signal in Figure 3.7. Considering the AM signal

$$u(t) = (1 + m\cos\Omega t)\cos\omega t \qquad \Omega \ll \omega \qquad m < 1$$

show that similar distortions are typical for other signals.

Solution Derivatives of $u(t)$ are as follows:

$$u' = -\omega(1 + m\cos\Omega t)\sin\omega t - m\Omega\sin\Omega t\cos\omega t$$

$$u'' = -\omega^2\left[1 + m\left(1 - \frac{\Omega^2}{\omega^2}\right)\cos\Omega t\right]\cos\omega t + 2m\Omega\omega\sin\Omega t\sin\omega t$$

and for $\Omega \ll \omega$, according to (3.27), we have

$$\omega(t)^2 = -\frac{u''}{u} \approx \omega^2 - \frac{2m\Omega\omega\sin\Omega t}{1 + m\cos\Omega t}\tan\omega t$$

Here, $\tan \omega t$ takes infinite values twice a period and makes distortions like in Figure 3.7. Amplitude expression is more complicated but includes the same tangent.

Problem 3.4

The TKA acceptable for narrowband signals is unsatisfactory for the wideband FM signal in Figure 3.9. Consider FM and AM signals for which the TKA is also unsatisfactory, whereas the AS gives reasonable APF.

Solution First we take quadratic FM:

$$u(t) = \sin\frac{t^3}{6} \qquad \omega(t) = \frac{t^2}{2} \qquad \omega(0) = 0$$

Around $t=0$ the signal and its derivatives are as follows: $u(t)=t^3/6$, $u'(t)=t^2/2$, $u''(t)=t$, $u'''(t) = 1$. Therefore, according to (3.28) and (3.29) we have:

$$\Psi(u) = (u')^2 - uu'' = \frac{t^4}{12} \qquad \Psi(u') = (u'')^2 - u'u''' = \frac{t^2}{2}$$

and the amplitude and frequency (3.30) are given by

$$a(t) = \frac{|t|^3}{6\sqrt{2}} \qquad \omega(t) = \frac{\sqrt{6}}{|t|}$$

So, for $t=0$, the TKA gives $\omega = \infty$ and $a=0$, though we expect $\omega=0$ and $a=1$. The TKA and AS amplitude and frequency are shown in Figure 3.12. Again, the AS meets our intuition, whereas the TKA gives a wrong answer.

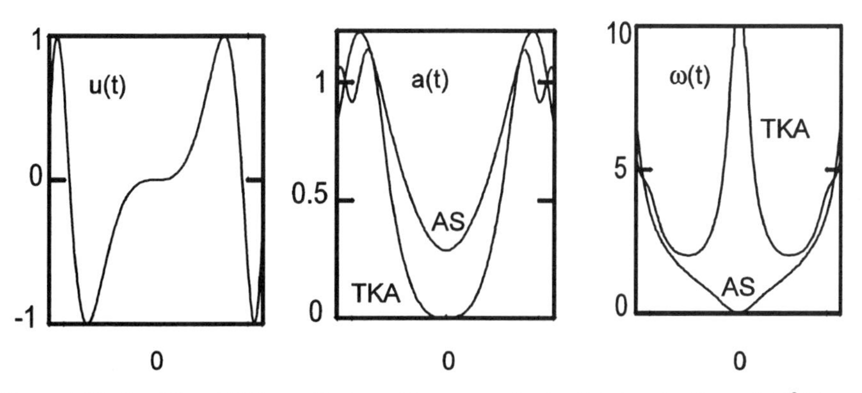

Figure 3.12 The AS and TKA amplitude and frequency for the FM signal $u(t) = \sin(t^3/6)$.

Another example is the AM signal

$$u(t) = a(t)\sin t \quad \text{where } a(t) = e^{t^2/6 - t^4/192}$$

This amplitude leads to zero cubic term in the power series

$$u(t) = t - \frac{19}{576}t^5 + \cdots \tag{3.47}$$

Therefore, as in Figure 3.9, the TKA amplitude is infinite and frequency is zero at $t = 0$. On the other hand, the AS amplitude is close to $a(t)$, and the frequency $\omega(t) \approx 1$ at $t = 0$ (Figure 3.13).

Problem 3.5

According to Problem 2.12, the AS amplitude defines signal energy as $E = \frac{1}{2}\int a^2 dt$. For the signal in Figure 3.13, show that its energy cannot be found from the TKA amplitude.

Solution Differentiating (3.47) and using (3.28) through (3.30), we find $a^2(t) \sim 1/t^2$ for $t \ll 1$. This amplitude is not integrable, and energy is undefined.

Problem 3.6

Shekel approximates a signal $u(t)$ with a sinusoid $a\cos(\omega t + \Phi)$ *at a given instant* (Section 3.5.3). Consider the approximation over a finite interval T for reducing the error

$$\epsilon = \int_{t-T/2}^{t+T/2} [u(t) - a\cos(\omega t + \Phi)]^2\, dt = \min \tag{3.48}$$

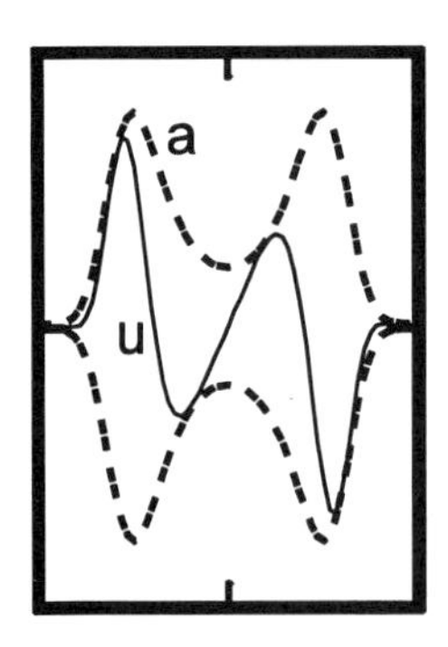

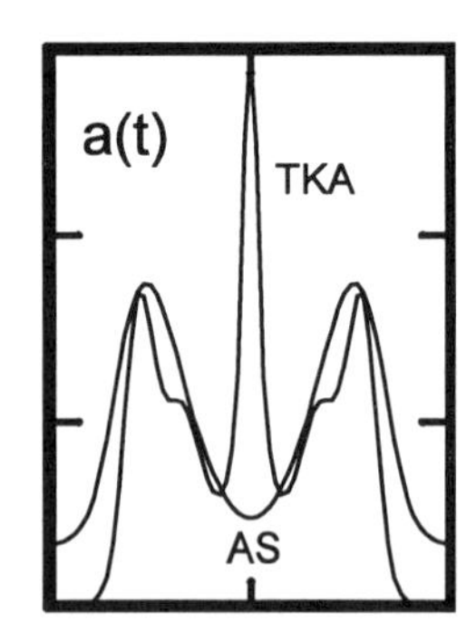

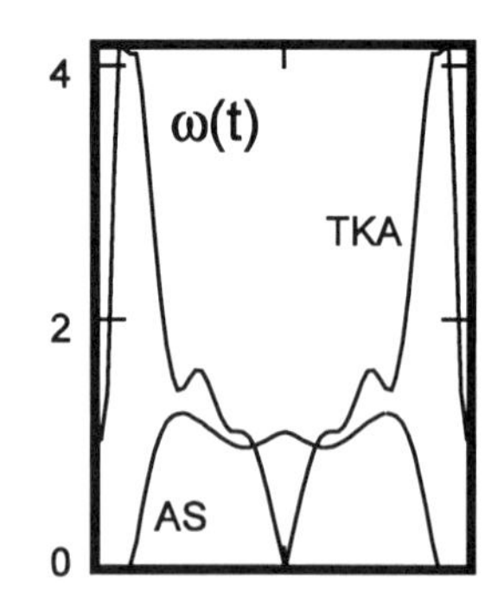

Figure 3.13 The AS and TKA amplitude and frequency for the wideband AM signal $u(t) = a(t)\sin t$. The amplitude $a(t) = e^{t^2/6 - t^4/192}$ is shown with dotted lines, and the AS amplitude is similar.

Why are the APFs resulting from this approximation also unsatisfactory?

Solution Denoting

$$E_t = \int_{t-T/2}^{t+T/2} u^2(t)\,dt \qquad U_t(\omega) = \int_{t-T/2}^{t+T/2} u(t)e^{-i\omega t}\,dt$$

and neglecting small oscillating addends at 2ω, we rewrite (3.48) as follows:

$$\epsilon = E_t - 2a\{\cos\Phi\cdot \mathrm{Re}\,U_t(\omega) + \sin\Phi\cdot \mathrm{Im}\,U_t(\omega)\} + \frac{a^2T}{2} \tag{3.49}$$

To define the APF, we minimize (3.49) with respect to a, Φ, and ω. The minimization can be done in any order, and differentiating with respect to Φ, we first find $\tan\Phi = \mathrm{Im}\,U_t(\omega)/\mathrm{Re}\,U_t(\omega)$. Substituting into (3.49), we then find

$$\min_{\Phi} \epsilon = E_t - 2a|U_t(\omega)| + \frac{a^2T}{2}$$

and minimizing with respect to amplitude, we have

$$\min_{\Phi, a} \epsilon = E_t - \frac{2}{T}|U_t(\omega)|^2$$

So, for a given t, the frequency of a signal is that for which the spectrum $|U_t(\omega)|^2$ is maximal. This approach looks reasonable, but it leads to irrelevant answers. For chirp signals, as shown in Figure 2.2, the spectrum often has *two equal maximums*, and it is unclear which one defines the signal frequency.

Problem 3.7

Show that the windows (3.39) are not unique for exact reconstruction from the semilocal spectrum (3.38). What condition is to be met in general?

Solution Substituting (3.35) and (3.38) into (3.42), we have

$$\frac{1}{2\pi}\int_{-\infty}^{\infty} W_t(\omega)e^{i\omega t}\,d\omega = \frac{w(t)}{2\pi}\int_{-\infty}^{\infty}\int_{-\infty}^{\infty} K_t(\omega)\,G_t(\tau)e^{-i\omega\tau + i\theta_t''\tau^2/2}\,d\tau\,d\omega$$

$$= w(t)\int_{-\infty}^{\infty} k_t(-\tau)\,G_t(\tau)\,e^{i\theta_t''\tau^2/2}\,d\tau$$

and exact reconstruction is achieved if

$$\int_{-\infty}^{\infty} k_t(-\tau)G_t(\tau)e^{i\theta_t''\tau^2/2}\,d\tau = 1$$

Here, $k_t(\tau)$ is the impulse response of the filter $K_t(\omega)$. In general, the integral depends on θ_t'', which should be compensated in $G_t(\tau)$. In (3.39), the compensation results in the factor $\sqrt{1+\gamma_t}$. So, various windows (and receivers) may provide exact reconstruction.

References

[1] Vakman, D. E., "Do We Know What Are the Instantaneous Frequency and Instantaneous Amplitude of a Signal?" translated in *Radio Eng. and Electron. Phys.*, Vol. 21, 1976, pp. 95–100.

[2] Wainstein, L. A., and D. E. Vakman, *Frequency Separation in Theory of Oscillations and Waves*, Moscow: Nauka-Press, 1983 (in Russian).

[3] Frenks, L. E., *Signal Theory*, Englewood Cliffs, NJ: Prentice Hall, 1969.

[4] Key, E. L., E. N. Fowle, and R. D. Haggarty, "A Method for Designing Signals of Large Time-Bandwidth Product," in *1961 IRE International Convention Record*, pt. 4, pp. 146–154.

[5] Vakman, D. E., *Sophisticated Signals and the Uncertainty Principle in Radar*, New York: Springer Verlag, 1968.

[6] Rytov, S. M., "On Paradoxes Related to Spectral Analysis," *Soviet Physics Successes*, Vol. 29, No. 1/2, 1946 (in Russian).

[7] Fink, L. M., *Signals, Interference, Mistakes*, Moscow: Radio and Communications, 1984 (in Russian).

[8] Vakman, D., "On the Analytic Signal, the Teager—Kaiser Energy Algorithm, and Other Methods for Defining Amplitude and Frequency," *Trans. IEEE Signal Processing*, Vol. 44, 1996, pp. 791–797.

[9] Andronov, A. A., S. E. Khaikin, and A. A. Vitt, *Theory of Oscillations*, Oxford, U.K.: Pergamon, 1966.

[10] Bogoliuboff, N. N., and Y. A. Mitropolskii, *Asymptotic Methods in the Theory of Nonlinear Oscillations*, New York: Gordon and Breach, 1961.

[11] Hafner, E., "The Effects of Noise in Oscillators," *Proc. IEEE*, Vol. 54, 1966, pp. 179–198.

[12] Hayashi, C., *Nonlinear Oscillations in Physical Systems*, New York: McGraw-Hill, 1964.

[13] Malahov, A. N., *Fluctuations in Auto-Oscillation Systems*, Moscow: Nauka-Press, 1968 (in Russian).

[14] Minorsky, N., *Nonlinear Oscillations*, Princeton, NJ: D. Van Nostrand, 1962.

[15] Rytov, S. M., *Introduction to Statistical Radio-Physics*, Moscow: Nauka-Press, 1976 (in Russian).

[16] Shekel, J., "Instantaneous Frequency," *Proc. of the IRE*, Vol. 41, 1953, p. 548.

[17] Kaiser, J. F., "On a Simple Algorithm to Calculate the 'Energy' of a Signal," *Proc. IEEE ICSSP '90,* Albuquerque, NM, 1990, pp. 381–384.

[18] Maragos, P., J. F. Kaiser, and T. F. Quatieri, "On Separating Amplitude from Frequency Modulation Using Energy Operators," in *Proc. IEEE ICSSP '92,* Vol. 2, San Francisco, CA, 1992, pp. 1–4.

[19] Bovik, A. C., P. Maragos, and T. F. Quatieri, "AM—FM Energy Detection and Separation in Noise Using Multiband Energy Operators," *IEEE Trans. Signal Processing,* Vol. 41, 1993, pp. 3245–3265.

[20] Vakman, D., L. Cohen, and L. B. White, "The Instantaneous Spectral Band," *Proc. IEEE-SP Symp. on Time—Freq. and Time—Scale Analysis,* Philadelphia, PA, 1994, pp. 28–31.

[21] Cohen, L., *Time-Frequency Analysis,* Englewood Cliffs, NJ: Prentice Hall, 1995.

[22] Lusternik, L. A., and V. I. Sobolev, *Elements of Functional Analysis,* Moscow: Nauka-Press, 1968 (in Russian).

4

The Analytic Signal in Radio Devices

In the beginning were physicists. They realized the nature of electromagnetic waves, determined their velocity (the light speed), and discovered that the waves are radiated from electric sparks and detected with a coherer containing a metal powder (H. Hertz, 1888). Besides, physicists discovered the electron (J. J. Thomson, 1897) and invented the electron tube (L. deForest, 1906). Without all that, radio-engineering would be impossible.

Engineers used the electromagnetic waves to transmit information in free space. A number of methods for modulation, detection, and transformation of signals have been suggested since the 1910s, and the most fruitful ones were selected for practical devices. Apparently, the most important radio-engineering methods were developed before 1936, when the first frequency modulation system was accomplished by E. Armstrong. It was 10 years before the analytic signal was suggested by Gabor.

All these methods operate with amplitudes and frequencies of signals. However, engineers did not think about analytic signals or the HT. They invented lowpass and bandpass amplifiers, mixers, detectors, modulators, and so on. Nevertheless, the amazing fact is that all practical radio devices were in accordance with the AS. So, the modulated frequency in Armstrong's system was the AS frequency, and the amplitude extracted with detectors was the AS amplitude. We will show it for common radio devices.

At the same time, physicists developed oscillation and wave theory, optics, lasers, and so forth. Even after 1946, however, the AS rarely occurs in works of physics, though it is preferable for those fields also. So, engineering practice has selected the best way while scientists were searching it.

4.1 Detection

4.1.1 Synchronous Detector

As illustrated in Figure 4.1, the synchronous detector extracts the inphase $x(t)$ and the quadrature $y(t)$ components (also called the *quadratures*) from a modulated signal

$$u(t) = a(t)\cos[\omega_0 t + \Phi(t)] = x(t)\cos\omega_0 t - y(t)\sin\omega_0 t$$

Each channel contains a multiplier and a lowpass filter. Reference signals $2\cos\omega_0 t$ and $2\sin\omega_0 t$ shifted by $-\pi/2$ are applied to multipliers, and in the upper channel, multiplication results in

$$\begin{aligned} u(t)\cdot 2\cos\omega_0 t &= [x(t)\cos\omega_0 t - y(t)\sin\omega_0 t]2\cos\omega_0 t \\ &= x(t) + x(t)\cos 2\omega_0 t - y(t)\sin 2\omega_0 t \end{aligned}$$

Then a lowpass filter separates the quadrature $x(t)$ from high-frequency oscillations at $2\omega_0$. In the same way, the lower channel extracts $y(t)$.

Clearly, the detector does not use the HT, AS, or any other specification of quadratures. Besides, its output signal depends on the filter employed. Even the ideal filter separates signals if their spectra are not overlapped, however. Therefore, the spectra $X(\omega)$ and $X(\omega - 2\omega_0)$ *must be apart* as shown in Figure 4.1, or equivalently, $X(\omega)$ and $Y(\omega)$ must be below the carrier frequency ω_0. In Chapter 2, we have seen, however, that $x(t)$ and $y(t)$ are then the quadratures of the AS (Bedrosian's theorem). Therefore, ignoring technical imperfections, we conclude that the detector extracts the AS quadratures because spectra are not overlapped.

For slow varying quadratures dependent on εt, we have almost the same. A signal band is narrow, of about $\varepsilon\omega_0$, and spectra are overlapped at a small level

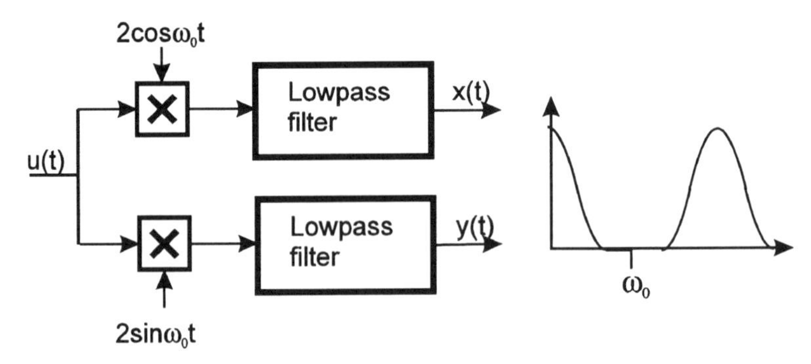

Figure 4.1 Synchronous detector.

ε^{r+1} (see Section 2.3). If we disregard this error, the output signals are the AS quadratures.

4.1.2 Quadratic and Linear Detectors

The quadratic detector consists of a nonlinear (quadratic) device and a lowpass filter as shown in Figure 4.2(a). Obviously, the input of the filter is

$$\begin{aligned} u(t)^2 &= [x(t)\cos\omega_0 t - y(t)\sin\omega_0 t]^2 \\ &= \frac{x^2+y^2}{2} + \frac{x^2-y^2}{2}\cos 2\omega_0 t - xy\sin 2\omega_0 t \end{aligned}$$

and the lowpass filter separates the low-frequency intensity

$$I(t) = \frac{x(t)^2 + y(t)^2}{2} = \frac{a(t)^2}{2}$$

from high-frequency oscillations at $2\omega_0$.

Again, spectra must not be overlapped. The spectrum of $x(t)^2$, for example, is the convolution $X(\omega) * X(\omega)$, which indicates that the frequency range is twice as large as the $X(\omega)$. Therefore, since spectra of $x(t)^2$ and $y(t)^2$ must be below ω_0, the $X(\omega)$ and $Y(\omega)$ must be below $\omega_0/2$. Then the $a(t)^2$ extracted is the AS amplitude squared.

A linear detector (rectifier) produces $|u(t)|$ instead of $u(t)^2$, and higher harmonics arise:

$$\begin{aligned} |u(t)| = a(t)|\cos\phi(t)| &= \frac{2}{\pi}a(t) + \frac{4}{3\pi}a(t)\cos 2\phi(t) \\ &\quad - \frac{4}{15\pi}a(t)\cos 4\phi(t) + \cdots \end{aligned}$$

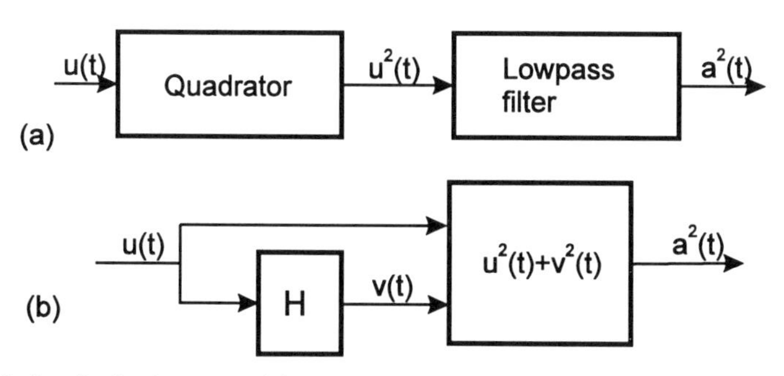

Figure 4.2 Quadratic detectors: (a) common circuit and (b) equivalent circuit with Hilbert transformer.

where Fourier series for $|\cos\phi|$ is used. As before, a lowpass filter separates the AS amplitude if spectra are not overlapped. Strictly speaking, the amplitude $a(t) = \sqrt{x(t)^2 + y(t)^2}$ is *not* band-limited even if the quadratures are. Therefore, the linear detector is not so rigorous as a quadratic one. For slow modulation, however, this distinction is unimportant.

4.2 Frequency Conversion and Multiplication

4.2.1 Frequency Conversion

The frequency converter (mixer) invented in 1918 is employed in all superheterodyne receivers. It transforms input frequencies to a fixed intermediate frequency. Amplification in the receiver is more effective at this fixed frequency since the amplifier should not be tunable.

The frequency converter also consists of a multiplier and a filter. However, a *bandpass* filter is used, and a reference frequency ω_1 differs from the signal frequency (Figure 4.3(a)). If the input signal is a sinusoid $\cos\omega t$, the multiplier makes two sinusoids:

$$2\cos\omega t\cos\omega_1 t = \cos(\omega - \omega_1)t + \cos(\omega + \omega_1)t$$

and a bandpass filter separates one of them, with plus or minus, which distinguishes the upward and downward conversions.

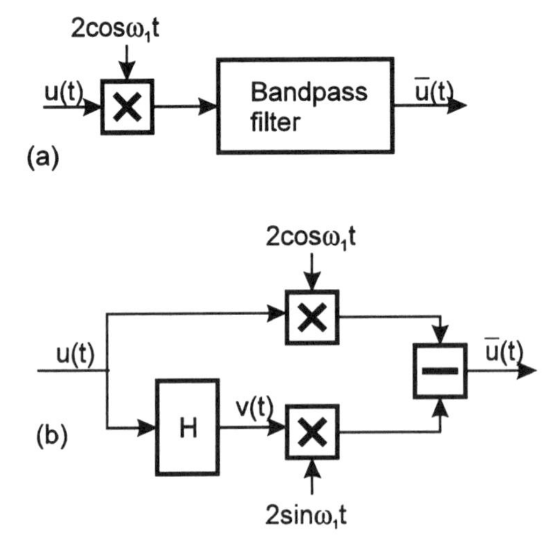

Figure 4.3 (a) Common and (b) balance frequency converters.

In general, a signal consists of many harmonics, and each spectral frequency is shifted by $\pm\omega_1$. Then the output signal is as follows:

$$\begin{aligned}\tilde{u}(t) &= \frac{1}{\pi}\int_0^\infty [U_c(\omega)\cos(\omega \pm \omega_1)t + U_s(\omega)\sin(\omega \pm \omega_1)t]\, d\omega \\ &= \frac{\cos\omega_1 t}{\pi}\int_0^\infty [U_c(\omega)\cos\omega t + U_s(\omega)\sin\omega t]\, d\omega \\ &\mp \frac{\sin\omega_1 t}{\pi}\int_0^\infty [U_c(\omega)\sin\omega t - U_s(\omega)\cos\omega t]\, d\omega \\ &= u(t)\cos\omega_1 t \mp v(t)\sin\omega_1 t \end{aligned} \tag{4.1}$$

Here, $u(t)$ is the input signal in the form (2.10), whereas $v(t)$ is its HT (2.11). Note that the HT comes now from the converter design and has no relation to the APF definition.

Modulation is to be conserved in a converter, however, and the output instantaneous frequency shifted by $\pm\omega_1$ should follow the input frequency. So, the converted signal must be as follows:

$$\tilde{u}(t) = a(t)\cos[\phi(t) \pm \omega_1 t] = a(t)\cos\phi(t)\cos\omega_1 t \mp a(t)\sin\phi(t)\sin\omega_1 t$$

Comparing with (4.1), we see that $u(t) = a(t)\cos\phi(t)$ and $v(t) = a(t)\sin\phi(t)$. Therefore, since $v(t)$ is the HT, *input modulation* must be accorded to the AS. Otherwise, modulation is distorted, and the converter cannot be used in a receiver.

Equation (4.1) shows also that the same conversion is achievable with the device in Figure 4.3(b) containing the Hilbert transformer H. This device is known as the *balance mixer*. It is often used in ultra-high-frequency receivers, and the transformer (phase-shifter) is commonly made with a quarter-wave line.

Now let us consider the output AS. If the reference frequency ω_1 is lower than the input spectrum of $u(t)$, applying the HT to (4.1) and using Bedrosian's theorem (2.22), we take out the low-frequency factors $\cos\omega_1 t$ and $\sin\omega_1 t$ and obtain

$$\tilde{v}(t) = \mathcal{H}[\tilde{u}(t)] = \mathcal{H}[u]\cos\omega_1 t \mp \mathcal{H}[v]\sin\omega_1 t = v\cos\omega_1 t \pm u\sin\omega_1 t$$

where we have used that $\mathcal{H}[u] = v$ and $\mathcal{H}[v] = -u$ (see Problem 2.1). Finally, combining with (4.1), we find the output AS in the form:

$$\begin{aligned}\tilde{w}(t) &= \tilde{u} + i\tilde{v} = u(\cos\omega_1 t \pm i\sin\omega_1 t) + v(i\cos\omega_1 t \mp \sin\omega_1 t) \\ &= (u + iv)e^{\pm i\omega_1 t} = w(t)e^{\pm i\omega_1 t}\end{aligned}$$

where the signs $\pm$ denote the upward and downward conversions. Similarly, if the spectrum of $u(t)$ is lower than ω_1, we take out $u(t)$ and $v(t)$ and obtain

$$\tilde{w}(t) = \begin{cases} w(t)e^{i\omega_1 t} & \text{for upward conversion} \\ w^*(t)e^{i\omega_1 t} & \text{for downward conversion} \end{cases}$$

In the second case, the mixer inverts the instantaneous frequency: rising frequency becomes falling and vice versa. This inversion is important for some applications.

In general, two arbitrary signals $u_1(t)$ and $u_2(t)$ are converted. If a *low-frequency* signal is $u_1(t)$, the resulting AS is as follows:

$$\tilde{w}(t) = \begin{cases} w_1(t)w_2(t) & \text{for upward conversion} \\ w_1^*(t)w_2(t) & \text{for downward conversion} \end{cases} \tag{4.2}$$

Clearly, the spectra of $u_1(t)$ and $u_2(t)$ must not be overlapped. Otherwise, the upper and lower components cannot be separated with a filter, and significant distortion appears.

4.2.2 Frequency Multiplication

The frequency multiplier increases the carrier frequency and broadens the spectrum. The simplest multiplier is similar to the quadratic detector in Figure 4.2(a), but the second harmonic is extracted with a *bandpass* filter. In general, the signal $u(t) = a(t)\cos\phi(t)$ is raised to the n-th power and the n-th harmonic is extracted from $u^n(t) = a^n(t)\cos^n\phi(t)$. Then the output signal takes the form $\tilde{u}(t) = a^n(t)\cos n\phi(t)$, and according to Problem 2.5, the associated AS transformation is as follows:

$$\tilde{w}(t) = w(t)^n \tag{4.3}$$

Again, nonoverlapping spectra are needed for frequency separation. More details are discussed in Problem 4.2.

4.3 Frequency Detection

As illustrated in Figure 4.4, the frequency detector contains a linear filter $K(\omega)$ that transforms the instantaneous input frequency $\omega(t)$ into amplitude. Then the amplitude is detected. Frequency-amplitude conversion can be written as

$$\tilde{w}(t) = K[\omega(t)]w(t) \tag{4.4}$$

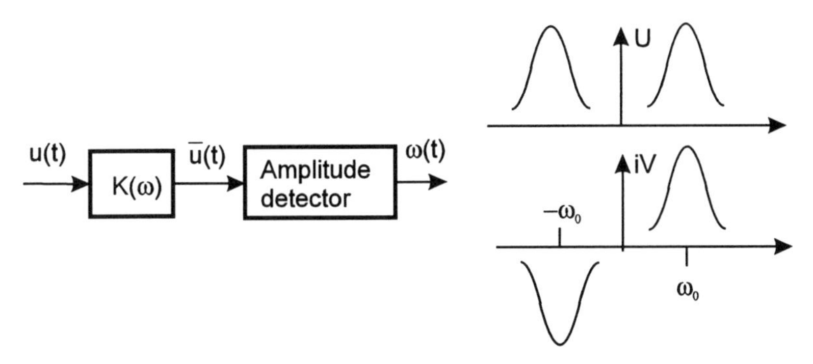

Figure 4.4 Frequency detector.

In early detectors, the conversion was made with a simple resonant circuit tuned to a linear slope of its frequency characteristic. Then, following the modulus $|K(\omega(t))|$, the output amplitude was linear with frequency.

The frequency response $K(\omega)$ is a function of a *spectral* frequency ω, but not an *instantaneous* frequency $\omega(t)$. Therefore, frequency detection assumes the local frequency notion where spectral and instantaneous frequencies are the same, and (4.4) should be justified. We will derive (4.4) carefully and show its relation to the AS. This relation is more complicated than for preceding devices, but spectral nonoverlapping is also necessary.

4.3.1 Relation to the AS

For a linear filter, the output real signal $\tilde{u}(t)$ is given by

$$L\left(\frac{d}{dt}\right)\tilde{u}(t) = u(t) = a(t)\cos[\omega_0 t + \Phi(t)] \quad \text{where } L\left(\frac{d}{dt}\right) = \sum_n c_n \frac{d^n}{dt^n} \tag{4.5}$$

This equation is for real signals, and the amplitude and frequency are ambiguous. Therefore, (4.4) cannot be deduced from (4.5) directly. However, it is derivable (see below) from the analogous *complex* equation

$$L\left(\frac{d}{dt}\right)\tilde{w}(t) = w(t) = a(t)e^{i[\omega_0 t + \Phi(t)]} \tag{4.6}$$

or, since $\tilde{w} = \tilde{u} + i\tilde{v}$, from the additional equation for imaginary signals

$$L\left(\frac{d}{dt}\right)\tilde{v}(t) = v(t) = a(t)\sin[\omega_0 t + \Phi(t)] \tag{4.7}$$

Let us show that (4.5) and (4.7) are interrelated with HT.

The initial physical equation is (4.5), and (4.7) must be deduced from it. This should be done with an operator $\mathcal{A}$ that transforms both sides as follows:

$$\begin{gathered}\mathcal{A}\left\{L\left(\frac{d}{dt}\right)\tilde{u}(t)\right\} = L\left(\frac{d}{dt}\right)\tilde{v}(t)\\ \mathcal{A}\{a(t)\cos[\omega_0 t + \Phi(t)]\} = a(t)\sin[\omega_0 t + \Phi(t)]\end{gathered} \tag{4.8}$$

The first line shows that $\mathcal{A}$ is a linear operator commutative with differentiation. Therefore, it is a stationary filter, and its frequency characteristic $A(\omega)$ is to be found from the second line.

Let us introduce the complex envelope $z(t) = a(t)e^{i\Phi(t)}$ and its spectrum $Z(\omega)$. Then the spectra of $u(t) = a(t)\cos[\omega_0 t + \Phi(t)]$ and $v(t) = a(t)\sin[\omega_0 t + \Phi(t)]$ are as follows:

$$U(\omega) = \frac{Z(\omega - \omega_0) + Z^*(-\omega - \omega_0)}{2}$$

$$V(\omega) = \frac{Z(\omega - \omega_0) - Z^*(-\omega - \omega_0)}{2i}$$

and from the second line (4.8), we find the filter's characteristic as follows:

$$A(\omega) = \frac{V(\omega)}{U(\omega)} = -i\frac{Z(\omega - \omega_0) - Z^*(-\omega - \omega_0)}{Z(\omega - \omega_0) + Z^*(-\omega - \omega_0)} \tag{4.9}$$

In general, $A(\omega)$ depends on $Z(\omega)$, and the signal-dependent operator $\mathcal{A}$ is not linear. Then the two lines in (4.8) are incompatible. However, this dependence cancels if spectra $Z(\omega - \omega_0)$ and $Z^*(-\omega - \omega_0)$ do not overlap each other as in Figure 4.4. Then (4.9) turns into the HT:

$$A(\omega) = \begin{cases} -i & \text{for } \omega > 0 \\ i & \text{for } \omega < 0 \end{cases}$$

So, (4.5) can be transformed into (4.7) for nonoverlapping spectra only, and in (4.6), the $\tilde{w}(t)$ must be the AS.

4.3.2 Carson-Fry Method

Now we derive (4.4) from the complex equation (4.6). Its general solution can be written as

$$\tilde{w}(t) = \frac{1}{2\pi}\int_0^\infty K(\omega)W(\omega)e^{i\omega t}\,d\omega \tag{4.10}$$

where $W(\omega)$ is the spectrum of the input AS $w(t) = a(t)e^{i\phi(t)}$ restricted to $\omega > 0$, and $K(\omega) = 1/L(i\omega)$. Expanding $K(\omega)$ around the instantaneous frequency $\omega(t)$, we obtain the series (see Problem 4.5 for details)

$$\tilde{w}(t) = K[\omega(t)]a(t)e^{i\phi(t)} - iK'[\omega(t)]a'(t)e^{i\phi(t)} - \frac{K''[\omega(t)]}{2}[a''(t) + i\omega'(t)a(t)]e^{i\phi(t)} + \cdots \tag{4.11}$$

This series was introduced by Carson and Fry in 1937 [1], long before the AS. Its first term is the frequency-amplitude conversion (4.4). The other corrections depend on amplitude and frequency rates and vanish for slow modulation. We see that the Carson-Fry method is also based on the AS.

Two errors characterize this method. Equation (4.6) is limited by spectral overlapping. According to Section 2.3, if the amplitude and phase dependent on slow time εt are differentiable r times, the error of equation is ε^{r+1}. On the other hand, the error of the series (4.11) is ε^{n+1} if we take n terms. To avoid *over-precision* when solution is more precise than initial equation, we should not take $n > r$. This limits the number of terms in (4.11) and shows that the Carson-Fry method is asymptotic.

4.4 Frequency Modulation

4.4.1 Roder's Vector Diagrams

An elegant approach to modulation was introduced by Roder in 1931 [2]. Representing each harmonic component of a real modulated signal

$$u(t) = \sum_k a_k \cos(\omega_k t + \Phi_k) \tag{4.12}$$

as a vector of length a_k rotating with the angular velocity ω_k, Roder *defined* the amplitude and phase of a signal as the instantaneous length and angle of the resulting vector. This approach is equivalent to the AS. Indeed, for a harmonic signal $a_k \cos(\omega_k t + \Phi_k)$, the rotating vector represents the complex exponent $a_k e^{i(\omega_k t + \Phi_k)}$, and a real signal (4.12) is supplemented with its HT in the imaginary part

$$v(t) = \sum_k a_k \sin(\omega_k t + \Phi_k) \tag{4.13}$$

The simplest AM signal consists of a carrier frequency ω_0 and two side frequencies $\omega_0 + \Omega$ and $\omega_0 - \Omega$ (usually $\Omega \ll \omega_0$):

$$u(t) = (1+m\cos\Omega t)\cos\omega_0 t = \cos\omega_0 t + \frac{m}{2}\cos(\omega_0+\Omega)t + \frac{m}{2}\cos(\omega_0-\Omega)t \tag{4.14}$$

In Figure 4.5, it is depicted as a sum of three rotating vectors. If we stop the carrier vector, the side vectors are rotating with angular velocities Ω and $-\Omega$. Their sum is varying in length (amplitude), but not in the angle (phase).

The most important Roder's idea was that frequency modulation can be represented in the same way. If the phases of side vectors are shifted by $\pi/2$, as shown in the lower diagram, the resulting vector is varying in the angle rather than in length, achieving frequency modulation.

That was a new and unexpected idea. AM and FM were considered as different phenomena achievable with dissimilar methods. Roder first understood that AM can be transformed into FM. He also replaced the local instantaneous frequency understandable at that time by the global AS frequency.

4.4.2 Armstrong's Modulator

Five years later, in 1936, E. Armstrong [3] created the first successful FM system for radio communications, and his modulator was completely accorded with the Roder's idea (and the AS). This modulator is shown in Figure 4.6.

The carrier signal of a high frequency ω_0 feeds two tubes of a balance modulator. The tubes are connected in contraphases, and for equal gains, the

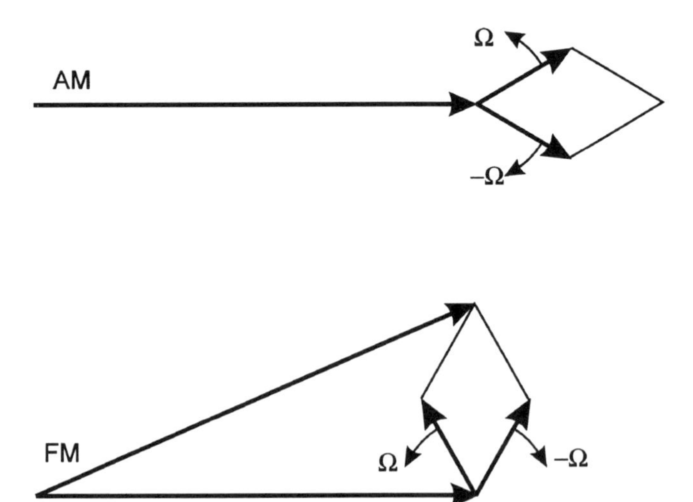

Figure 4.5 Roder's diagrams for AM and FM signals.

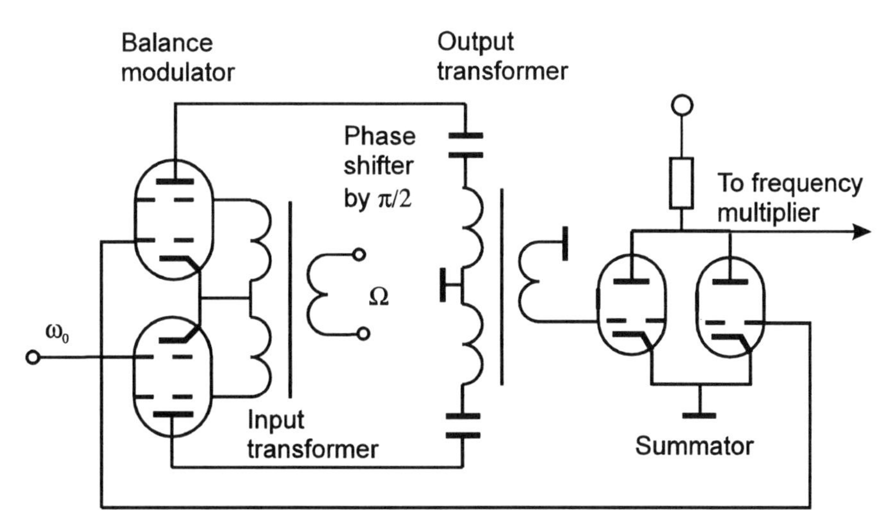

Figure 4.6 Armstrong's frequency modulator.

output current (in the transformer) is zero. The modulating signal of a low frequency Ω feeds screen grids of the tubes. It varies their gains so that the carrier component is compensated out and only the side components remain:

$$\begin{aligned}(1 + m\cos\Omega t)\cos\omega_0 t - (1 - m\cos\Omega t)\cos\omega_0 t \\ = m\cos(\omega_0 + \Omega)t + m\cos(\omega_0 - \Omega)t\end{aligned}$$

Then, the resonant circuit formed with the output transformer and additional capacitors shifts the phases by $\pi/2$. The resonant circuit is tuned to ω_0. Therefore, the current in the transformer is in phase with a signal, but the voltage on the secondary winding is shifted by $\pi/2$. Finally, shifted side components are added to the carrier in a summator (two amplifiers with a common load), which creates the FM signal in full agreement with Roder's diagrams. So, the instantaneous modulated frequency was the frequency of the AS.

Figure 4.5 shows also that FM is accompanied with parasitic AM, and the greater the side components, the greater the amplitude distortion. For small distortion, the FM index should be small, and modulation should be *narrowband.* On the other hand, a local frequency detector used in the Armstrong's system was for *wideband* modulation. Armstrong reconciled the global modulator and the local detector with a frequency multiplier, which increases the deviation and index by 500 times. Therefore, the asymptotic relationship between local and global notions was also used in Armstrong's system. Let us recall again that it was 10 years before the AS was introduced, which gave engineers their due for comprehension of such delicate features of signals.

4.5 Supplementary Problems

Problem 4.1

In differential geometry, the envelope $a(t)$ of a family of curves $u(t, \theta)$ dependent on the additional parameter θ is defined by

$$a = u \quad \text{for } \frac{\partial u}{\partial \theta} = 0 \tag{4.15}$$

On the other hand, we have defined the envelope (amplitude) of a signal through the AS. Considering output signals $\tilde{u}(t, \omega_1)$ of the mixer as a family with the reference frequency ω_1 as a parameter, show that both definitions are identical.

Solution According to (4.1), for $\theta = \omega_1 t$, we have from (4.15)

$$a(t) = u(t)\cos\theta - v(t)\sin\theta \quad \text{for} \quad u(t)\sin\theta + v(t)\cos\theta = 0$$

Then, eliminating θ, we come to the AS amplitude $a(t) = \sqrt{u^2 + v^2}$. So, for a varying reference frequency, the envelope of the output signals is the AS amplitude.

Problem 4.2

Show that frequency multiplication broadens the band of a signal by n times. Also, show that the factor of frequency multiplication is limited by spectral overlapping. Use the following properties of entire functions [4]:

- The function $z(t)$ of a complex t is an *entire function of the exponential type* α (or belongs to the class B_α) if it is analytic at every finite point and obeys the inequality

$$|z(t)| \leq C\exp(\alpha|t|) \tag{4.16}$$

  that turns into equality for $|t| \to \infty$ in some ray on the complex t-plane (for a proper constant C).
- According to the Wiener-Paley theorem, $z(t)$ belongs to B_α if and only if its spectrum is band-limited to $-\alpha \leq \omega \leq \alpha$.

Solution Let the input bandwidth be 2α. Then, the input AS is $w(t) = z(t)e^{i\omega_0 t}$ for $\omega_0 > \alpha$, and its complex envelope $z(t)$ belongs to B_α. According to (4.3), the output AS takes the form

$$w^n = z^n e^{in\omega_0 t}$$

and because of (4.16)

$$|z^n| \leq C^n \exp(n\alpha|t|)$$

So, z^n belongs to $B_{n\alpha}$ and its spectrum is limited to the band $|\omega| \leq n\alpha$.

In a frequency multiplier, a bandpass filter extracts the n-th harmonic from

$$u^n(t) = a^n \cos^n[\omega_0 t + \Phi] = \frac{a^n}{2^n}\left[e^{i(\omega_0 t + \Phi)} + e^{-i(\omega_0 t + \Phi)}\right]^n$$
$$= \frac{a^n}{2^n}\sum_{k=0}^{n} C_n^k e^{i(n-2k)(\omega_0 t + \Phi)} = \frac{a^n}{2^n}\sum_{k=0}^{n} C_n^k \cos[(n-2k)(\omega_0 t + \Phi)]$$

where C_n^k are binomial coefficients, and we have taken into account that $u^n(t)$ is a real function. So, the filter separates the oscillation $a^n(t)\cos n[\omega_0 t + \Phi(t)]$ at $n\omega_0$ from the oscillation at $(n-2)\omega_0$. If the band $n\alpha > \omega_0$, however, their spectra are overlapped, and the filter cannot work properly. Therefore, depending on the input band α, the factor of multiplication is limited by $n < \omega_0/\alpha$.

Problem 4.3

The frequency response of the filter *matched to the signal* $u(t)$ is the complex conjugate spectrum of its AS $w(t)$, so that $\hat{K}(\omega) = W^*(\omega)$. Find the complex impulse response of the matched filter and show that its output signal is the autocorrelation function of the input AS.

Solution The impulse response is Fourier transform of $W^*(\omega)$:

$$\hat{k}(t) = \frac{1}{2\pi}\int_0^\infty W^*(\omega)e^{i\omega t}\, d\omega = w^*(-t)$$

and the output AS given by the third formula (2.27) is as follows:

$$\tilde{w}(t) = \frac{1}{2}\int_{-\infty}^{\infty} w(s)w^*(s-t)\, ds$$

that is, the autocorrelation function of $w(t)$.

Problem 4.4

Let the reference signal in the mixer be $e^{i(\omega_0 t - \mu t^2)}$, and let its output filter be matched to this chirp signal. Show that the mixer produces the spectrum of the input signal.

Solution The mixer multiplies complex signals according to (4.2), and the matched filter produces the convolution as in Problem 4.3. Therefore, we have

$$\tilde{w}(t) = \int_{-\infty}^{\infty} w(s) e^{i[\omega_0 s - \mu s^2]} e^{-i[\omega_0 (s-t) - \mu (s-t)^2]} \, ds$$

$$= e^{i(\omega_0 t + \mu t^2)} \int_{-\infty}^{\infty} w(s) e^{-i2\mu ts} \, ds = e^{i(\omega_0 t + \mu t^2)} W(2\mu t)$$

Thus, the output amplitude $|\tilde{w}(t)|$ follows the modulus $|W(2\mu t)|$ of the input spectrum, and the real-time spectral analyzer is obtained. The frequency scale depends on the rate μ in the reference chirp signal.

Problem 4.5

Derive the Carson-Fry series (4.11).

Solution We substitute the Taylor series expansion

$$K(\omega) = \sum_{n=0}^{\infty} \frac{K^{(n)}(\omega(t))}{n!} [\omega - \omega(t)]^n$$

where $K^{(n)}$ mean derivatives with respect to ω, into (4.10) to obtain

$$\tilde{w}(t) = \sum_{n=0}^{\infty} \frac{K^{(n)}(\omega(t))}{2\pi n!} \int_0^{\infty} [\omega - \omega(t)]^n W(\omega) e^{i\omega t} \, d\omega$$

$$= \sum_{n=0}^{\infty} \frac{(-i)^n}{n!} K^{(n)}(\omega(t)) \mathcal{D}_n w(t)$$

Here, the operators $\mathcal{D}_n$ are given by

$$\mathcal{D}_n w(t) = \frac{i^n}{2\pi} \int_0^{\infty} [\omega - \omega(t)]^n W(\omega) e^{i\omega t} \, d\omega \qquad (4.17)$$

Further, differentiating with respect to t, we find

$$\frac{d}{dt}\mathcal{D}_n w = \frac{i^n}{2\pi}\int_0^{\infty}\{-n\omega'(t)[\omega-\omega(t)]^{n-1}+i\omega[\omega-\omega(t)]^n\}\,W(\omega)\,e^{i\omega t}\,d\omega$$
$$= -in\omega'(t)\mathcal{D}_{n-1}w+\mathcal{D}_{n+1}w+i\omega(t)\mathcal{D}_n w$$

where we have used $\omega[\omega-\omega(t)]^n = [\omega-\omega(t)]^{n+1}+\omega(t)[\omega-\omega(t)]^n$. We come to the recurrent relation:

$$\mathcal{D}_{n+1}w = \left[\frac{d}{dt}-i\omega(t)\right]\mathcal{D}_n w + in\omega'(t)\mathcal{D}_{n-1}w$$

From (4.17), we also have $D_0 w = w$ and $D_1 w = w' - i\omega(t)w$, and (4.11) results by induction.

Problem 4.6

AM and FM are both doubleband in the sense that two symmetrical bands exist around the carrier, and a total bandwidth is at least two times wider than the bandwidth of the modulating signal. Describe a method for singleband modulation.

Solution To get a singleband signal, we move a low-frequency modulating spectrum to a high-carrier frequency. The mixer for upward conversion does so, and the balance mixer in Figure 4.3(b) is also known as the Hartley singleband modulator [5].

References

[1] Carson, J. R., and T. C. Fry, "Variable Frequency Electric Circuit Theory with Application to the Theory of Frequency Modulation," *Bell System Tech. J.*, Vol. 16, 1937, pp. 513–540.

[2] Roder, H., "Amplitude, Phase, and Frequency Modulation," *Proc. of the IRE*, Vol. 19, 1931, pp. 2145–2176.

[3] Armstrong, E. H., "A Method for Reducing Disturbances and Radio Signaling by a System of Frequency Modulation," *Proc. of the IRE*, Vol. 24, 1936, pp. 689–740.

[4] Khurgin, Ya. I., and V. P. Yakovlev, *Finite Functions in Physics and Engineering*, Moscow: Nauka Press, 1971 (in Russian).

[5] Frenks, L. E., *Signal Theory*, Englewood Cliffs, NJ: Prentice Hall, 1969.

5

Random Signals and Noise

The amplitude and phase of a noise were introduced by Rice [1] in 1944, but then it was shown that they are the same as for the AS [2]. The AS widely accepted in noise theory provides elegant correlation functions for complex random signals and simplifies many problems. In this chapter, we briefly discuss correlation properties and explore two particular problems connected with the local and global instantaneous frequency of random signals.

The first problem is random frequency modulation with a noise. We will show that spectra of modulated signals are different for slow and fast modulation. Slow FM creates a spectrum that follows the probability distribution of a modulating noise. This agrees with the local frequency notion when the instantaneous and spectral frequencies are the same. However, fast FM provides the Lorentzian spectrum typical for diffusion. So, modulation and diffusion are the phenomena pertaining to the local and global frequency notions.

The second problem is the most practical in the book. Precise measurement of frequency stability is very important for engineering, physics, and radioastronomy. However, up-to-date measurements are based on the local frequency notion, and considerable measuring errors arise for short-term stability. We will show that the global AS provides a great advantage of accuracy over local methods. We will also show that the statistical theory of instability should be revised. A paradoxical and important fact is that frequency variations within a short time may and should be measured during a much longer time.

5.1 Correlation Properties of Noise

Basic definitions related to real random processes are given in Problem 5.7. Here, we consider interrelations with complex random processes based on the AS.

5.1.1 Spectral Densities and Correlation Functions

If $u(t)$ is a real stationary random process of zero mean (noise), its spectral density and autocorrelation function are defined by

$$S_u(\omega) = \overline{|U(\omega)|^2} = \overline{\left|\int_{-\infty}^{\infty} u(t)e^{-i\omega t}\,dt\right|^2} \tag{5.1}$$

$$R_u(\tau) = \overline{u(t)u(t-\tau)} = \frac{1}{\pi}\int_0^{\infty} S_u(\omega)\cos\omega\tau\,d\omega \tag{5.2}$$

where an overbar now denotes statistical averaging (over the ensemble of realizations). Note that $S_u(\omega)$ is an even function. Due to stationarity, R_u depends on τ but not t. Note also that additional normalization is needed in (5.1) as discussed in Problem 5.7.

Because of time-stationarity (property 4 AS in Section 2.2), applying the HT to $u(t)$, we create the Hilbert-conjugate real stationary random process $v(t)$ and the complex random AS $w(t) = u(t) + iv(t)$. According to (2.14), spectra of $u(t)$ and $v(t)$ differ in a phase only. Therefore, their spectral densities and correlation functions are identical:

$$S_v(\omega) = \overline{|V(\omega)|^2} = \overline{|U(\omega)|^2} = S_u(\omega) \tag{5.3}$$

$$R_v(\tau) = \overline{v(t)v(t-\tau)} = R_u(\tau) \tag{5.4}$$

The cross-spectral density for v and u is defined as $S_{vu}(\omega) = \overline{V(\omega)U^*(\omega)}$ and results from (2.14) as follows:

$$S_{vu}(\omega) = -i\,\mathrm{sgn}(\omega)\overline{|U(\omega)|^2} = \begin{cases} -iS_u(\omega) & \text{for } \omega > 0 \\ iS_u(\omega) & \text{for } \omega < 0 \end{cases} \tag{5.5}$$

Therefore, $R_{vu}(\tau)$ is the HT of $R_u(\tau)$:

$$R_{vu}(\tau) = \overline{v(t)u(t-\tau)} = \frac{1}{\pi}\int_{-\infty}^{\infty} \frac{R_u(s)}{\tau - s}\,ds = \frac{1}{\pi}\int_0^{\infty} S_u(\omega)\sin\omega\tau\,d\omega \tag{5.6}$$

Finally, the correlation function for the AS $w(t) = u + iv$ is given by

$$\begin{aligned} R_w(\tau) &= \frac{1}{2}\overline{w(t)w^*(t-\tau)} = \frac{1}{2}\overline{[u(t) + iv(t)][u(t-\tau) - iv(t-\tau)]} \\ &= \frac{1}{2}\overline{u(t)u(t-\tau)} + \frac{i}{2}\overline{v(t)u(t-\tau)} - \frac{i}{2}\overline{u(t)v(t-\tau)} + \frac{1}{2}\overline{v(t)v(t-\tau)} \\ &= R_u(\tau) + iR_{vu}(\tau) = \frac{1}{\pi}\int_0^{\infty} S_u(\omega)e^{i\omega\tau}\,d\omega \end{aligned} \tag{5.7}$$

and therefore

$$S_w(\omega) = S_u(\omega) + iS_{vu}(\omega) \tag{5.8}$$

The factor 1/2 in correlation functions of complex processes simplifies many relations. Fourier integrals in (5.6) and (5.7) are obtained in the same way as in (2.11) and (2.12) but for the even spectrum $|U(\omega)|^2 = S_u(\omega)$. We have also taken into account that $R_v(\tau) = R_u(\tau)$ and that

$$R_{uv}(\tau) = \overline{u(t)v(t-\tau)} = \overline{u(t'+\tau)v(t')} = R_{vu}(-\tau) = -R_{vu}(\tau)$$

where (5.6) is used again. Thus, the correlation functions R_u, R_{vu}, and R_w are interrelated in the same way as u, v, and w.

Following the same development as in (5.7), we find that

$$\overline{w(t)w(t-\tau)} = R_u(\tau) + iR_{vu}(\tau) + iR_{uv}(\tau) - R_v(\tau) = 0 \tag{5.9}$$

Differentiating (5.9), we also find that derivatives are uncorrelated, in particular

$$\overline{w'(t)w(t-\tau)} = 0 \quad \text{and} \quad \overline{w'(t)w'(t-\tau)} = 0 \tag{5.10}$$

This indicates that the AS $w(t)$ and its shifted version $w(t-\tau)$ plus their derivatives are uncorrelated at any τ, though $w(t)$ and $w^*(t-\tau)$ define the correlation function R_w of a complex noise.

Intensity (variance) of a real noise $u(t)$ can be found from either its correlation function $R_u(\tau)$ or the correlation function $R_w(\tau)$ for the AS. In fact, for $\tau = 0$, from (5.2) and (5.7), we have

$$\overline{u^2(t)} = \frac{1}{2}\overline{|w(t)|^2} = R_u(0) = R_w(0) = \frac{1}{\pi}\int_0^\infty S_u(\omega)\,d\omega \tag{5.11}$$

5.1.2 Mean Spectral Frequency and Bandwidth

The mean frequency of a noise and its bandwidth are defined by

$$\overline{\omega} = \frac{1}{\pi R_w(0)}\int_0^\infty \omega S_u(\omega)\,d\omega \tag{5.12}$$

$$\Delta\omega^2 = \frac{1}{\pi R_w(0)}\int_0^\infty (\omega - \overline{\omega})^2 S_u(\omega)\,d\omega = \overline{\omega^2} - \overline{\omega}^2 \tag{5.13}$$

where

$$\overline{\omega^2} = \frac{1}{\pi R_w(0)}\int_0^\infty \omega^2 S_u(\omega)\,d\omega$$

From (5.7) we have $\overline{\omega} = -iR'_w(0)/R_w(0)$, $\overline{\omega^2} = -R''_w(0)/R_w(0)$, where the derivatives are taken with respect to τ and then set $\tau = 0$. Also, rewriting (5.7) in the form

$$2R_w(\tau_1 + \tau_2) = \overline{w(t + \tau_1)w^*(t - \tau_2)}$$

differentiating with respect to τ_1 and τ_2 and then setting $\tau_1 = \tau_2 = 0$, we find that $2R'_w(0) = \overline{w'(t)w^*(t)}$ and $2R''_w(0) = -\overline{|w'(t)|^2}$. Therefore,

$$\overline{\omega} = -i\frac{R'_w(0)}{R_w(0)} = -i\frac{\overline{w'(t)w^*(t)}}{\overline{|w(t)|^2}} \tag{5.14}$$

$$\Delta\omega^2 = -\frac{R''_w(0)}{R_w(0)} + \left[\frac{R'_w(0)}{R_w(0)}\right]^2 = \frac{\overline{|w'(t)|^2}}{\overline{|w(t)|^2}} + \left[\frac{\overline{w'(t)w^*(t)}}{\overline{|w(t)|^2}}\right]^2 \tag{5.15}$$

5.1.3 Relation to Amplitude and Frequency

We now introduce the amplitude and frequency of a noise by setting $w(t) = a(t)e^{i\phi(t)}$ and $\omega(t) = \phi'(t)$. Since $w' = (a' + i\omega a)e^{i\phi}$, multiplying by $w^* = ae^{-i\phi}$, we find

$$\begin{aligned} w'(t)w^*(t) &= a'(t)a(t) + i\omega(t)a^2(t) \\ |w'(t)|^2 &= a'(t)^2 + \omega^2(t)a^2(t) \end{aligned} \tag{5.16}$$

Then, for the normalized noise with $\overline{|w(t)|^2} = \overline{a^2(t)} = 1$, averaging (5.16) and comparing with (5.14) and (5.15), we finally have

$$\overline{\omega} = -i\overline{a'(t)a(t)} + \overline{\omega(t)a^2(t)} = \overline{\omega(t)a^2(t)} \tag{5.17}$$

$$\Delta\omega^2 = \overline{\left[\frac{a'(t)^2}{a^2(t)} + \omega^2(t)\right]a^2(t)} - [\overline{\omega(t)a^2(t)}]^2 \tag{5.18}$$

Since $\overline{\omega}$ is real, we have concluded that $\overline{a'(t)a(t)} = 0$, so that the noise amplitude and its derivative are uncorrelated. The relations obtained generalize Fink's formulas (2.30) for statistical averaging instead of time and frequency averaging. They show that, when averaging over realizations, the mean spectral frequency and bandwidth depend on the instantaneous frequency and amplitude of a noise.

5.1.4 Complex Envelope and Quadratures

We also introduce the complex envelope of a noise as follows:

$$z(t) = a(t)e^{i\Phi(t)} = x(t) + iy(t) = w(t)e^{-i\omega_0 t} \tag{5.19}$$

Then, according to (5.7), its correlation function and spectral density are given by

$$R_z(\tau) = \frac{1}{2}\overline{z(t)z^*(t-\tau)} = \frac{1}{2}\overline{w(t)w^*(t-\tau)}e^{-i\omega_0\tau} = R_w(\tau)e^{-i\omega_0\tau} \tag{5.20}$$

$$\begin{aligned} S_z(\omega) &= \int_{-\infty}^{\infty} R_z(\tau)e^{-i\omega\tau}\,d\tau = \int_{-\infty}^{\infty} R_w(\tau)e^{-i(\omega+\omega_0)\tau}\,d\tau \\ &= S_w(\omega+\omega_0) = \begin{cases} 2S_u(\omega+\omega_0) & \text{for } \omega > -\omega_0 \\ 0 & \text{for } \omega < -\omega_0 \end{cases} \end{aligned} \tag{5.21}$$

From (5.9), we also have $\overline{z(t)z(t-\tau)} = 0$ and $\overline{z^*(t)z^*(t-\tau)} = 0$. Therefore, for the quadrature $x = (z+z^*)/2$, we find the following correlation function:

$$\begin{aligned} R_x(\tau) &= \overline{x(t)x(t-\tau)} = \frac{1}{4}\overline{[z(t)+z^*(t)][z(t-\tau)+z^*(t-\tau)]} \\ &= \frac{1}{4}\overline{z(t)z^*(t-\tau)} + \frac{1}{4}\overline{z^*(t)z(t-\tau)} = \frac{1}{2}[R_z(\tau)+R_z^*(\tau)] = \mathrm{Re}[R_z(\tau)] \end{aligned}$$

Since $y = (z - z^*)/2i$, similar development for $R_y(\tau)$ and $R_{yx}(\tau)$ leads to correlation functions

$$\begin{aligned} R_x(\tau) &= R_y(\tau) = \mathrm{Re}[R_z(\tau)] \\ R_{yx}(\tau) &= -R_{xy}(\tau) = \mathrm{Im}[R_z(\tau)] \end{aligned} \tag{5.22}$$

In the same way as (5.21), we also obtain spectral densities of quadratures:

$$\begin{aligned} S_x(\omega) &= S_y(\omega) = \frac{1}{2}\left[S_w(\omega_0-\omega) + S_w(\omega_0+\omega)\right] \\ S_{yx}(\omega) &= -S_{xy}(\omega) = \frac{i}{2}\left[S_w(\omega_0-\omega) - S_w(\omega_0+\omega)\right] \\ &\text{where } S_w(\omega) = \begin{cases} 2S_u(\omega) & \text{for } \omega > 0 \\ 0 & \text{for } \omega < 0 \end{cases} \end{aligned} \tag{5.23}$$

is the spectral density of a complex noise $w(t)$.

5.2 Random Frequency Modulation

Random FM shows an interesting interrelation between local and global frequency notions. For slow modulation, instantaneous and spectral frequencies are the same, and the spectral density follows the probability of a modulating noise. Fast modulation, however, results in the Lorentzian spectrum of diffusion that cannot be explained within a local approach.

Let a signal be frequency-modulated with a real normal noise $n(t)$ of the Gaussian probability density

$$p(n) = \frac{1}{\sqrt{2\pi}\sigma} e^{-n^2/2\sigma^2} \qquad \sigma^2 = \overline{n^2} \tag{5.24}$$

and let the spectral density of the noise be $N(\omega)$. If the carrier frequency ω_0 is sufficiently high, the AS and complex envelope of the modulated signal are given by

$$w(t) = e^{i[\omega_0 t + \Phi(t)]} \qquad z(t) = e^{i\Phi(t)} \qquad \Phi(t) = \int_{t_0}^{t} n(t')\, dt' \tag{5.25}$$

where the lower limit defines the initial phase and may be arbitrary.

Within a framework of the local frequency notion, the signal spectrum should follow the probability density (5.24) of the modulating frequency $\omega(t) = \omega_0 + n(t)$. Indeed, the probability density defines the mean time that dwells at a given frequency ω. The longer this time, the larger value of the spectral density $S_w(\omega)$. However, this customary reasoning is true for slow modulation only. For fast modulation, spectral and instantaneous frequencies are different, and the spectrum is typical for diffusion.

According to (5.20) and (5.25), the correlation function of the complex envelope $z(t)$ is given by

$$R_z(\tau) = \frac{1}{2}\overline{z(t)z^*(t-\tau)} = \frac{1}{2}\overline{e^{i\Delta\Phi(\tau)}} \tag{5.26}$$

where

$$\Delta\Phi(\tau) = \Phi(t) - \Phi(t-\tau) = \int_{t-\tau}^{t} n(t')\, dt' \tag{5.27}$$

is the random phase increment accumulated in a time τ. Then the spectral density is defined as

$$S_z(\omega) = \int_{-\infty}^{\infty} R_z(\tau) e^{-i\omega\tau}\, d\tau = \frac{1}{2}\int_{-\infty}^{\infty} \overline{e^{i\Delta\Phi(\tau)}} e^{-i\omega\tau}\, d\tau \tag{5.28}$$

5.2.1 The Mean-Square Phase Increment and the Spectrum

The phase increment (5.27) is a superposition of Gaussian noise values $n(t')$ and is a Gaussian variable as well (see Problem 5.7). Its variance relates to the correlation function $R_n(\tau)$ of a noise and to its spectral density $N(\omega)$ as follows:

$$\begin{aligned}\Delta^2(\tau) &= \overline{\Delta\Phi^2(\tau)} = \int_{t-\tau}^{t}\int_{t-\tau}^{t} \overline{n(s_1)n(s_2)}\, ds_1\, ds_2 \\ &= \int_0^{\tau}\int_0^{\tau} R_n(s_1 - s_2)\, ds_1\, ds_2 \qquad (5.29) \\ &= \frac{1}{2\pi}\int_0^{\tau}\int_0^{\tau} ds_1\, ds_2 \int_{-\infty}^{\infty} N(\omega) e^{i\omega(s_1-s_2)}\, d\omega \\ &= \frac{4}{\pi}\int_0^{\infty} N(\omega)\frac{\sin^2(\omega\tau/2)}{\omega^2}\, d\omega \qquad (5.30)\end{aligned}$$

where we substituted $R_n(s_1 - s_2)$ with the inverse Fourier transform of the spectral density $N(\omega)$. Also, the order of integration was changed.

So $\Delta\Phi$ is a Gaussian variable of a variance Δ^2. Therefore, as discussed in Problem 5.7, the mean value $\overline{e^{i\Delta\Phi}}$ can be found from its probability density function:

$$R_z(\tau) = \frac{1}{2}\overline{e^{i\Delta\Phi}} = \frac{1}{2\sqrt{2\pi}\,\Delta}\int_{-\infty}^{\infty} e^{i\Delta\Phi} e^{-\Delta\Phi^2/2\Delta^2}\, d\Delta\Phi = \frac{1}{2}e^{-\Delta^2(\tau)/2} \qquad (5.31)$$

Note that in view of the identity $a^2x^2 + bx = (ax + b/2a)^2 - b^2/4a^2$, the last integral is taken as follows:

$$\begin{aligned}\int_{-\infty}^{\infty} e^{-(a^2x^2+bx)}\, dx &= e^{b^2/4a^2}\int_{-\infty}^{\infty} e^{-(ax+b/2a)^2}\, dx \\ &= e^{b^2/4a^2}\int_{-\infty}^{\infty} e^{-z^2}\frac{dz}{a} = \frac{\sqrt{\pi}}{a} e^{b^2/4a^2}\end{aligned}$$

where we set $x = \Delta\Phi$, $a = 1/\sqrt{2}\Delta$, and $b = -i$.

Finally, using (5.31), the spectrum (5.28) becomes

$$S_z(\omega) = \frac{1}{4}\int_{-\infty}^{\infty} e^{-\Delta^2(\tau)/2 - i\omega\tau}\, d\tau \qquad (5.32)$$

Hence, for Gaussian frequency modulation, the mean-square phase increment $\Delta^2(\tau)$ strictly defines the spectrum (5.32) of a random signal. In turn, according to (5.30), this increment depends on the spectral density $N(\omega)$ of a modulating noise.

5.2.2 Modulation and Diffusion

Let the modulating spectrum be uniform in a band Ω: $N(\omega) = N_0$ for $0 < \omega < \Omega$. Then (5.30) is as follows:

$$\Delta^2(\tau) = \frac{4N_0}{\pi}\int_0^{\Omega}\frac{\sin^2(\omega\tau/2)}{\omega^2}\,d\omega = N_0\tau f(\Omega\tau) \qquad (5.33)$$

and the function $f(\Omega\tau)$ is shown in Figure 5.1. The function linearly increases at $\Omega\tau \ll 1$ and approaches $f = 1$ at $\Omega\tau \gg 1$. So, we have

$$\Delta^2(\tau) = \begin{cases} \dfrac{N_0\Omega}{\pi}\tau^2 & \text{for } \Omega\tau \ll 1 \\ N_0\tau & \text{for } \Omega\tau \gg 1 \end{cases} \qquad (5.34)$$

These limit cases correspond to different phenomena—modulation and diffusion.

For a small τ, the noise $n(t)$ is strongly correlated, and its variations are small within τ. Therefore, the random phase increment (5.27) is linear with respect to τ, and its variance Δ^2 is quadratic like was given by the first line of (5.34). On the other hand, the random phase increment for a large τ contains many *independent* increments. Similar to the Brownian motion, this results in *phase diffusion.* Then the variance is linear, like was given by the second line of (5.34).

Let us calculate the spectra of modulation and diffusion. For a *narrowband* modulating noise, $N_0/\Omega \gg 1$, and the exponent in (5.32) approaches zero so fast that the first relation (5.34) is valid for the whole integral. Then the spectrum

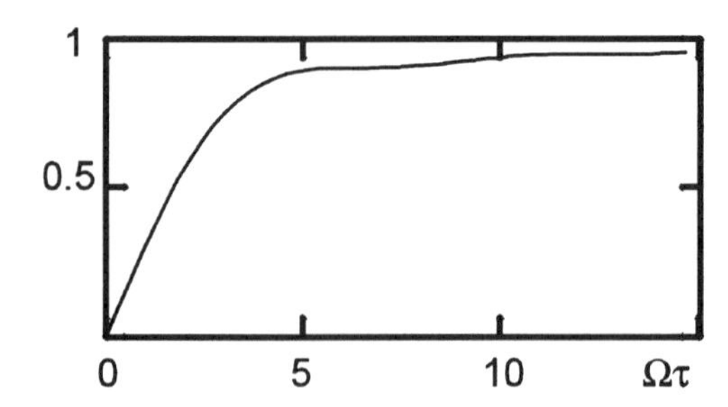

Figure 5.1 The function $f(\Omega\tau)$ defining the mean-square phase increment.

follows the Gaussian probability (5.24) of a modulating noise:

$$S_z(\omega) = \frac{1}{4}\int_{-\infty}^{\infty} e^{-\frac{N_0\Omega}{2\pi}\tau^2 - i\omega\tau}\,d\tau = \frac{\sqrt{\pi}}{2\sqrt{2}\sigma}e^{-\omega^2/2\sigma^2} \tag{5.35}$$

Here $\sigma^2 = N_0\Omega/\pi$ is the noise variance, and the integral is taken in the same way as (5.31). So, the spectrum $S_z(\omega)$ follows the probability of the modulating noise. This agrees with the *local* frequency notion: spectral and instantaneous frequencies are the same, and the spectral density follows the probability of the instantaneous frequency.

However, for a *wideband* (white) modulating noise, we set $\Omega = \infty$, and the second line (5.34) becomes exact. Then we obtain the Lorentzian spectrum of diffusion very different from (5.35):

$$S_z(\omega) = \int_0^{\infty} e^{-N_0\tau/2}\cos\omega\tau\,d\tau = \frac{N_0/2}{N_0^2 + 4\omega^2}. \tag{5.36}$$

Note that the second line in (5.34) is written for $\tau > 0$. In general, $\Delta^2(\tau) = N_0|\tau|$ is the even function. To avoid a negative τ, we use the trigonometric Fourier integral with $\cos\omega\tau$. Also, the integral is taken with a substitution $\cos\omega\tau = (e^{i\omega\tau} + e^{-i\omega\tau})/2$.

So, noise modulation and diffusion are interrelated phenomena. The wider a band of a modulating noise, the greater diffusion dominates over modulation. The local frequency notion interprets modulation only, but the global AS describes diffusion also.

5.3 Frequency Stability Measurement

To determine the frequency variation in a time T, we commonly measure frequencies at the beginning and end of T during a *short* time interval $\Delta T \ll T$. For the special case $\Delta T = T/2$, this results in the *Allan's frequency variance* widely used for stability estimation [4–6, 17]. This *local* approach assumes that the frequency can be measured during a short time ΔT with any desired precision. The shorter ΔT, however, the bigger the associated measuring errors in a noise.

The *global* approach to frequency measurements is very different [3]. Frequencies and their *short-term* variations are measured during a *long* time. We will show that this approach provides much higher precision for short-term instability. We will also show that statistical theory of instability developed by many authors [4–9] should be revised.

It is often assumed that, for narrowband quasiharmonic signals, the local approach is sufficient and the AS is needless. The problem of frequency measurements shows, however, that the AS is advantageous even for extremely narrowband signals of highly stable generators.

5.3.1 Local Measurements

Figure 5.2 shows frequency instability of high-precision quartz generators. The frequency of 5 MHz produced with a generator was converted to 10 kHz with a mixer, and duration ΔT of N periods (between zero-crossings) was measured with a time counter. Then the frequency was calculated as

$$\omega = \frac{2\pi N}{\Delta T} \tag{5.37}$$

and Allan's variance was plotted for each tested generator as a function of ΔT.

These results were obtained by Barber in 1971 [10], and he concluded that long-term instability (for $\Delta T > 1$ sec) is different for the generators, but short-term instability (for $\Delta T < 0.1$ sec) is identical for all of them and increases as ΔT^{-1}. However, the ΔT^{-1} dependence may be inherent to local measurements rather than to generators.

A weak noise $n(t)$ shifts the position of zero by $\delta t = n/a\omega$. For not too short ΔT, the random shifts $\delta t_1 = n_1/a\omega$ and $\delta t_2 = n_2/a\omega$ are independent at the beginning and the end. Then frequency errors and their variance are

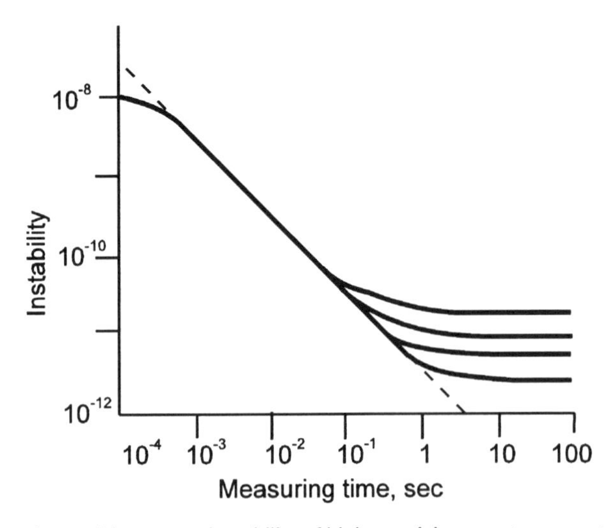

Figure 5.2 Experimental frequency instability of high-precision quartz generators.

as follows:

$$\delta\omega = \frac{2\pi N}{\Delta T - \delta t_1 + \delta t_2} - \frac{2\pi N}{\Delta T} \approx \frac{n_1 - n_2}{a\Delta T} \qquad \sigma_\omega^2 = \overline{\delta\omega^2} \approx \frac{2\,\overline{n^2}}{a^2 \Delta T^2} \tag{5.38}$$

For the 90-dB signal-to-noise ratio, relative standard deviation σ_ω/ω is also shown in Figure 5.2 (dotted straight line). Clearly, the short-term instability may be nonmeasurable under the noise (or discrete errors in a time counter equivalent to additive noise). For very short $\Delta T \sim 10^{-4}$ sec, correlation of noises n_1 and n_2 explains the divergence of curves in the left upper side. So, it can be assumed that only long-term instability for $\Delta T > 1$ sec is reliable in Barber's experiments, but the ΔT^{-1} dependence comes from measuring errors.

For many stable oscillators (atom standards, lasers, etc.), recent measurements show the same pattern [9,11,12]. In spite of the higher measuring accuracy, experimental points lie on straight lines like in Figure 5.2, and depending on environments, the critical ΔT may be of 1 hour instead of 1 sec.

For the *internal* generator noise integrated in the counter with a signal, the errors are slightly different. If the ΔT is not too short, many independent noise increments are accumulated in ΔT, and like for diffusion in (5.34), $\overline{n^2} \sim \Delta T$. Then, by virtue of (5.38), the ΔT^{-1} dependence is replaced by $\Delta T^{-1/2}$. Both dependencies are visible in practical measurements, but both may come from measuring errors.

Zero-crossing measurement is appealing in the sense of independence of phase (frequency) definition. Indeed, for $a(t) \neq 0$, any phase of a complex signal $a(t)e^{i\phi(t)}$ increases by 2π between zeros of $\cos\phi(t)$. However, noise distortions may be measured instead of generator instability.

5.3.2 Statistical Theory of Instability

Theory of instability explains the same experimental data in another way [4–6,9,13]. Based on the local frequency notion, the theory assumes that the *averaged* instantaneous frequency

$$\bar{\omega}(t) = \frac{1}{\Delta T}\int_{t-\Delta T}^{t} \omega(t)\,dt = \frac{1}{\Delta T}\left[\phi(t) - \phi(t - \Delta T)\right] \tag{5.39}$$

is the only measurable characteristic of instability. This averaged frequency is really measured with a time counter. Indeed, $2\pi N$ in (5.37) is the phase increment $\phi(t) - \phi(t - \Delta T)$ accumulated in ΔT between zeros. So, the theory describes the common but particular measuring method with zero-crossings.

The Allan's variance is constructed from the adjacent averaged frequencies and given by [17]

$$\bar{\sigma}_{\omega}^{2} = \overline{[\bar{\omega}(t+\Delta T) - \bar{\omega}(t)]^{2}} = \frac{1}{\Delta T^{2}}\overline{[\phi(t+\Delta T) - 2\phi(t) + \phi(t-\Delta T)]^{2}}$$

Since the time shift by $\pm\Delta T$ results in the factor $e^{\mp i\omega\Delta T}$ in the spectrum, this relation leads to

$$\begin{aligned} \bar{\sigma}_{\omega}^{2} &= \frac{1}{2\pi\Delta T^{2}}\int_{-\infty}^{\infty} S_{\phi}(\omega)(e^{-i\omega\Delta T} - 2 + e^{i\omega\Delta T})^{2}\, d\omega \\ &= \frac{16}{\pi\Delta T^{2}}\int_{0}^{\infty} S_{\phi}(\omega)\sin^{4}\left(\frac{\omega\Delta T}{2}\right) d\omega \end{aligned} \tag{5.40}$$

Here $S_{\phi}(\omega)$ is the spectral density of a random phase $\phi(t)$ defined by

$$S_{\phi}(\omega) = \int_{-\infty}^{\infty} R_{\phi}(\tau)e^{-i\omega\tau}\, d\tau \quad \text{where } R_{\phi}(\tau) = \overline{\phi(t)\phi(t-\tau)}$$

For the flicker spectral density $S_{\phi}(\omega) = B/\omega$, (5.40) gives $\bar{\sigma}_{\omega} \sim \Delta T^{-1}$, and for $S_{\phi}(\omega) = B/\omega^{2}$, it gives $\bar{\sigma}_{\omega} \sim \Delta T^{-1/2}$. So, the experimental data in Figure 5.2 can also be interpreted as generator instability under flicker-type random phase modulation. Since the flicker spectrum is singular at $\omega = 0$, the theory often uses complicated mathematical methods [8,14]. Because of this singularity, it has even been assumed that the classical statistical variance may be inapplicable to metrology [15].

Thus, the theory associates the experimental data with random flicker-type phase modulation. However, fundamental questions remain open:

- Does the experimental data really come from the additive noise or frequency modulation?
- Is the averaged frequency (5.39) the only measurable characteristic of stability?
- Can the flicker-type phase modulation be reasserted with direct measurements or the generator theory?

The global measuring method discussed below allows us to answer the questions. It eliminates additive noises, and *modulation instability* is measured only. Therefore, phase flicker modulation can be observed experimentally. Also, the instantaneous frequency of the AS is another measurable characteristic of instability. In Chapter 6, we will consider flicker modulation in generators and show that it produces frequency distortions very different from those assumed in the instability theory.

5.3.3 Global Measurements

Noise Fluctuations of the Analytic Signal Frequency

Let $w(t)$ be the AS of a real measured signal $u(t)$. According to (2.6), its instantaneous frequency is as follows:

$$\omega(t) = \mathrm{Im}\left[\frac{w'(t)}{w(t)}\right] \tag{5.41}$$

If a weak noise $n(t)$ is added to $u(t)$, frequency distortions appear. Random frequency fluctuations and their variance are given by

$$\delta\omega = \mathrm{Im}\left[\frac{w' + w'_n}{w + w_n} - \frac{w'}{w}\right] \approx \mathrm{Im}\left[\frac{w'_n w - w' w_n}{w^2}\right] \tag{5.42}$$

$$\sigma_\omega^2 = \overline{\mathrm{Im}^2\left[\frac{w'_n w - w' w_n}{w^2}\right]} = \frac{\overline{|w'_n w - w' w_n|^2}}{2|w|^4} \tag{5.43}$$

where the last relation in (5.43) is deduced in Problem 5.3. Here, w_n is the random AS associated with a noise n, whereas $w = a(t)e^{i\phi(t)}$ is a *deterministic* measured signal even if it contains random modulation (a sampled function). Therefore, averaging relates to w_n but not w. In (5.42), we neglected the second-order correction with respect to small w_n.

It is also shown in Problem 5.3 that (5.43) can be reduced to

$$\sigma_\omega^2 = \frac{\overline{n^2}}{a^2}\left[\Delta\omega^2 + [\omega(t) - \overline{\omega}]^2 + \frac{a'(t)^2}{a(t)^2}\right] \tag{5.44}$$

Here $\overline{\omega}$ and $\Delta\omega$ are the mean frequency and the effective bandwidth of a noise as defined in (5.12) and (5.13), whereas $a(t)$ and $\omega(t)$ are the amplitude and frequency of a measured signal. If the measured frequency is centered within a noiseband and the amplitude is constant, (5.44) takes a simple form

$$\sigma_\omega^2 = \frac{\overline{n^2}}{a^2}\Delta\omega^2 \tag{5.45}$$

and shows the fundamental distinction between local and global measurements.

Comparing (5.45) and (5.38), we see that when the interval ΔT is measured with a time counter, the noise is filtered in a band $\Delta\omega \sim 1/\Delta T$. In Barber's experiments, the narrowband noise was so weak for $\Delta T > 1$ sec that the errors were small, and long-term instability was measured properly. However, for $\Delta T < 0.1$ sec, the noiseband was so wide that frequency variations were covered with noise. For short-term instability, the measuring time ΔT is short, and the errors of the averaged frequency (5.39) are necessarily great. So, the dependence between ΔT and measuring accuracy is inherent to local measurements.

In the global AS method, narrowband filtering of a small $\Delta\omega$ is directly employed, and neighboring frequencies are measured with high precision independently of the interval ΔT between samples. Moreover, in contrast with local methods, precision is higher for short ΔT.

Filtering and Computer Measuring

We have seen in Section 2.4 that, when filtering, the output AS is a convolution of a real input signal $u(t)$ and a complex filter's response $\hat{k}(t)$:

$$w(t) = \int_{\infty}^{\infty} u(s)\hat{k}(t-s)\,ds \quad \text{and} \quad w'(t) = \int_{\infty}^{\infty} u(s)\hat{k}'(t-s)\,ds \tag{5.46}$$

Then the instantaneous frequency of a *filtered* signal is defined from (5.41). This is the basic algorithm of filtering and frequency measurement in the global method. Using a filter, we also avoid numerical or analog differentiation of $u(t)$ plus its HT. The method is more efficient for computer processing, and now we discuss some details of its realization.

Filter's Characteristic As shown in Section 2.4, the filter's response $K(\omega)$ must be band-limited to $\omega > 0$ for $\hat{k}(t)$ to be the AS. On the other hand, $\hat{k}(t)$ must be time-limited for the integration time in (5.46) to be finite. A compromise for these conflicting requirements is Gaussian characteristic

$$K(\omega) = e^{-(\omega-\omega_0)^2 T^2/2}$$

restricted to $-8/T < \omega - \omega_0 < 8/T$ and $\omega_0 > 8/T$. Then, the truncated values are of about 10^{-14}, and the filter's sidelobes (in a time domain) are of –280 dB.

Allowing wideband signals like chirps, we generally employ the characteristic

$$\begin{aligned} K(\omega) &= \frac{T}{\sqrt{2\pi}} \int_{\omega-\omega_0-\delta}^{\omega-\omega_0+\delta} e^{-s^2T^2/2}\,ds \\ &= \frac{T}{\sqrt{2\pi}} \left\{ \int_{-\infty}^{\omega-\omega_0+\delta} - \int_{-\infty}^{\omega-\omega_0-\delta} \right\} e^{-s^2T^2/2}\,ds \end{aligned} \tag{5.47}$$

which is the difference of two normal distributions. It turns into Gaussian characteristic for a small δ and approaches a rectangle (with a flat top and Gaussian edges) for a large δ. For better noise suppression, the filter should be tuned to the measured signal: its central frequency ω_0 and bandwidth δ are to be fitted to the first and second spectral moments of $u(t)$ (like in (5.12) and (5.13) with replacing $S_u(\omega)$ by $|U(\omega)|^2$). Other filters are also acceptable if they eliminate negative frequencies to a high degree.

Quantization Noise For computer processing, the input signal $u(t)$ must be sampled in the ADC and saved in a computer's memory. For a sampling period Δt, the m-bit ADC generates a white noise of quantization within the Nyquist band $0 < \omega < \Omega = \pi/\Delta t$. The noise is uniformly distributed within a step of quantization. For the signal amplitude fitted to the ADC range, its variance is as follows:

$$\frac{\overline{n^2}}{a^2} = \frac{1}{12} 2^{-2(m-1)}$$

Exceeding other fluctuations, this noise is a main source of errors. However, a digital filter of a narrowband $\Delta\omega$ reduces this variance (plus the variance of the internal generator noise) by a factor $\Delta\omega/\Omega$, and according to (5.45), the output frequency variance takes the form

$$\sigma_\omega^2 = \frac{1}{12} 2^{-2(m-1)} \frac{\Delta\omega}{\Omega} \Delta\omega^2 = \frac{1}{12\pi} 2^{-2(m-1)} \Delta t \Delta\omega^3 \qquad (5.48)$$

For the 14-bit ADC with the sampling period $\Delta t = 1$ ms and for the effective filter's band $\Delta\omega/2\pi = 1$ Hz, the errors (5.48) are of about 10^{-6} Hz. Therefore, at the 5-MHz nominal frequency, relative errors are less than 10^{-12}. Duration T of the filter's response (and time of a single measurement) is about 1 sec. Nevertheless, frequency variations are measurable for any time down to the sampling period of 10^{-3} sec.

Short-Term Errors In addition, long measuring intervals considerably overlap each other for close samples. Therefore, the noise accumulated in (5.46) is canceled, and short-term errors are additionally reduced by a factor $\Delta\omega\Delta t/2\pi$ (for every sampling period measurement). Therefore, the final frequency variance is as follows:

$$\sigma_\omega^2 = \frac{1}{24\pi^2} 2^{-2(m-1)} \Delta t^2 \Delta\omega^4 \qquad (5.49)$$

which provides the additional gain of about 40 dB. So, resulting relative errors approach 10^{-14}. In comparison with the errors in Figure 5.2 (of about 10^{-8} at

10^{-3} sec), the advantage of accuracy over the local method is 10^6. This great advantage is because of a long measuring time (1 sec instead of 1 ms) that comes from the global frequency notion. In fact, we have used the global idea of taking a long measuring time around each t.

Fast Fourier Transforms For sampled signals, the integrals (5.46) are replaced by sums, and these convolutions are easily calculated with *fast Fourier transforms* (FFT). Direct FFT defines the spectrum $U(\omega)$ of the input noisy signal $u(t)$, and after filtering, inverse FFT defines the output AS $w(t)$ and its derivative $w'(t)$. The FFT is a very effective programming technique for frequency measurements (see also Problem 5.5).

5.3.4 Simulation

To examine the accuracy of global measurements for short-time frequency variations, we simulated a signal with a linear frequency change of a given rate μ:

$$u(t) = a\cos[\omega_0 t + \Phi + \pi\mu(t - 4096)^2] + n(t) \qquad t = 0, 1, \ldots, 8191 \tag{5.50}$$

Here, ω_0 and Φ are the random initial frequency and phase. The ADC was simulated by taking the integer part of $u(t)$ and the amplitude $a = 8000$ was in accordance with the 14-bit ADC. The quantization noise is a fractional part of $u(t)$, and the internal generator noise $0 < n(t) < 1$ within a step of quantization was also added.

After quantization, the spectrum $U(\omega)$ of the signal (5.50) was computed with direct FFT. Then the first and second spectral moments were determined, and the filter's central frequency ω_0 and bandwidth δ were fitted to the moments. This defined the characteristic $K(\omega)$ in the form (5.47). Further, the products $K(\omega)U(\omega)$ and $i\omega K(\omega)U(\omega)$ defined the spectra of the output AS and its derivative, and the $w(t)$ and $w'(t)$ were calculated with inverse FFT. Finally, the array of measured frequencies $\omega(t)$ was found from (5.41). In spite of a long filter's response, the frequencies were computed for each sample.

For $\Delta T = 1$ and $\Delta T = 100$ sampling periods, frequency changes

$$\delta f = [\omega(t + \Delta T) - \omega(t)]/2\pi$$

were computed at 60 random instants within signal duration. According to frequency modulation in (5.50), they should equal μ and 100μ, respectively. For a comparison, Table 5.1 shows the mean δf and standard deviation for each series of 60 measurements. Particular results are also presented in Figures 5.3 through 5.5.

Since the sampling period $\Delta t = 1$ ms is taken as a time unit, all frequencies are given in kHz. For the given frequency rate μ, the figures illustrate tuning of the filter (graphs on the left) and random measuring errors for 1 and 100 sampling periods (*ER*1, *ER*100, graphs on the right), which are also presented in decibels with respect to 1 kHz.

At the 5-MHz nominal frequency, the rate $\mu = 10^{-10}$ kHz per sample makes relative frequency variations of about 10^{-14}/ms. As shown in Table 5.1, even such small frequency changes are measured with errors (standard deviation) of 5% or less. For the rate $\mu = 10^{-8}$ kHz per sample and relative variations of 10^{-12}/ms, measuring errors are about 0.05%. Both short-term absolute errors (standard deviations of δf_1) are less than 10^{-11} kHz. Graphs on the right in Figures 5.3 and 5.4 show also that, because of overlapping of measuring intervals, short-term errors *ER*1 are less than long-term errors *ER*100 by 40 dB. The ω-scale in the left graphs (8192 points/kHz) is defined by FFT.

Figure 5.5 and the third line in the Table 5.1 display the errors for a wideband chirp signal. Its spectrum approaches a rectangle and so does the

Table 5.1
Synoptical Simulation Results

Chirp Rate (kHz/ms)	Mean (δf_1) (kHz)	Mean (δf_{100}) (kHz)	Stdev (δf_1) (kHz)	Stdev (δf_{100}) (kHz)
$1 \cdot 10^{-10}$	$1.0032 \cdot 10^{-10}$	$9.980 \cdot 10^{-9}$	$3.8 \cdot 10^{-12}$	$2.9 \cdot 10^{-10}$
$1 \cdot 10^{-8}$	$9.9982 \cdot 10^{-9}$	$1.0000 \cdot 10^{-6}$	$3.9 \cdot 10^{-12}$	$3.8 \cdot 10^{-10}$
$1 \cdot 10^{-6}$	$9.9996 \cdot 10^{-7}$	$9.9994 \cdot 10^{-5}$	$5.9 \cdot 10^{-10}$	$4.8 \cdot 10^{-8}$

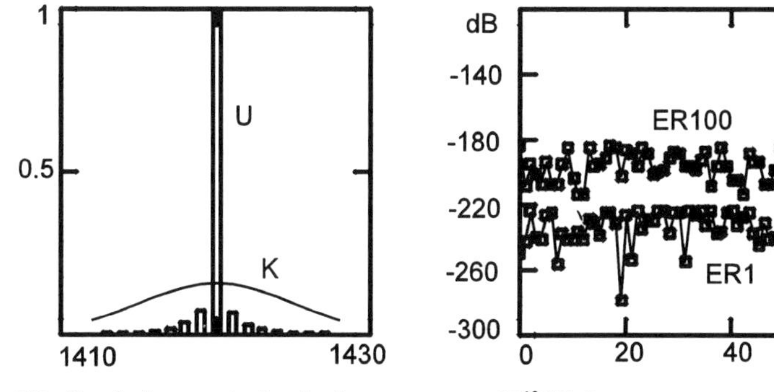

Figure 5.3 Simulating results for the frequency rate 10^{10} kHz/ms.

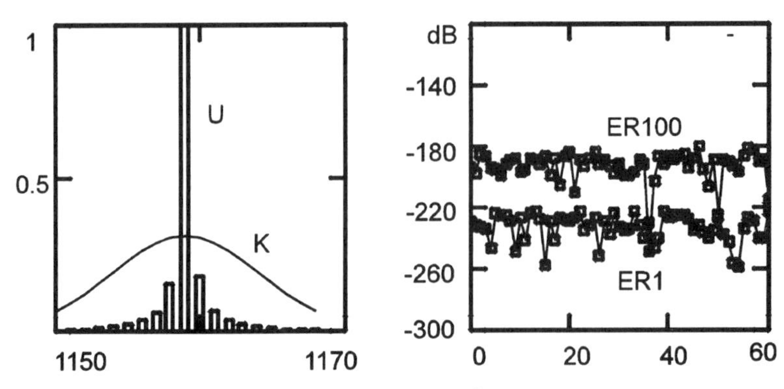

Figure 5.4 Simulating results for the frequency rate 10^8 kHz/ms.

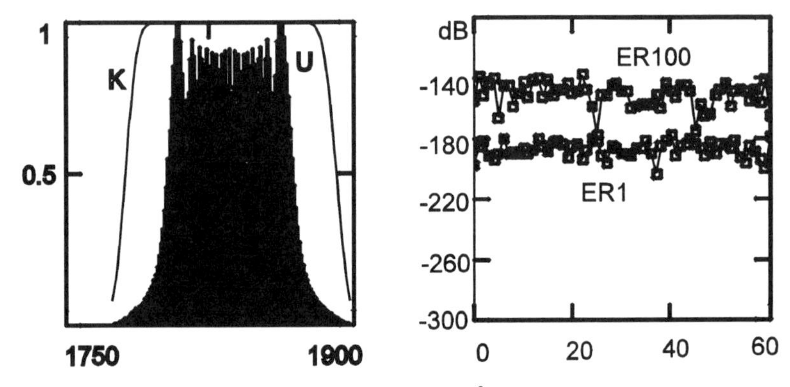

Figure 5.5 Simulating results for the frequency rate 10^6 kHz/ms.

filter's characteristic (graph on the left). Because of the wide filter's band, the accuracy is less than that in Figure 5.3, but still very high.

5.3.5 Discussion

Global frequency measurement based on the AS provides accuracy unattainable for common local methods, and additive noises are completely removed. Also, in contrast to common methods, accuracy is higher for short-time frequency changes.

For effective noise suppression, narrowband digital filtering is used. In spite of a long filter's response and long measuring time, short-time frequency variations are measurable. As a result, noise distortions are eliminated but *modulation* frequency variations caused by random or intentional perturbations can be measured with high precision.

Reducing additive noises to a high degree, the method verifies the statistical theory of instability. The theory assumes that flicker-type phase modulation exists in oscillators, but the same results may be caused with additive noise. Our method measures frequency (or phase) modulation, and flicker-type modulation will be observed if it exists. We believe, however, that for highly stable generators, up-to-date measurements are strongly disturbed with noise, and the actual short-term stability is much higher.

Some measuring methods have already used the AS. So, the method in [16] uses the two-channel synchronous detector and, as shown in Chapter 4, determines the AS frequency. In theory, this method is close to ours, but computer processing provides higher precision. Strict frequency definition given by the AS clears a way for this processing, and the highest precision is due to narrowband digital filtering stipulated by the global AS nature.

5.4 Supplementary Problems

Problem 5.1

Strictly speaking, Fourier integral (5.1) is divergent since the stationary noise $u(t)$ does not approach zero for $t \to \pm\infty$. Additional normalization for removing the divergence is discussed in Problem 5.7. On the other hand, the correlation function $R_u(\tau)$ is a decaying function, and the integral (5.2) converges. Using the HT in a time domain, derive (5.6) and (5.4) in another way.

Solution The cross-correlation function $R_{vu}(\tau)$ is given by

$$R_{vu}(\tau) = \overline{v(t)u(t-\tau)} = \frac{1}{\pi}\int_{-\infty}^{\infty} \frac{\overline{u(s)u(t-\tau)}}{t-s}\,ds$$
$$= \frac{1}{\pi}\int_{-\infty}^{\infty} \frac{R_u(s-t+\tau)}{t-s}\,ds = \frac{1}{\pi}\int_{-\infty}^{\infty} \frac{R_u(\sigma)}{\tau-\sigma}\,d\sigma = \mathcal{H}\left[R_u(\tau)\right]$$

that is, (5.6). In the same way, using Problem 2.1, we find

$$R_v(\tau) = \overline{v(t)v(t-\tau)} = \frac{1}{\pi}\int_{-\infty}^{\infty} \frac{\overline{u(s)v(t-\tau)}}{t-s}\,ds$$
$$= \mathcal{H}\left[R_{uv}(\tau)\right] = -\mathcal{H}\left[R_{vu}(\tau)\right] = -\mathcal{H}^2\left[R_u(\tau)\right] = R_u(\tau)$$

that is, (5.4). This rigorous approach provides all correlation properties from Section 5.1.

Problem 5.2

In Section 5.2, we have considered frequency modulation with a normal noise of the uniform spectral density $N(\omega) = N_0$ at $0 \le \omega \le \Omega$. Analyzing the modulating noise of Lorentzian spectral density

$$N(\omega) = \frac{N_0}{1 + \omega^2/\Omega^2},$$

so that

$$R_n(\tau) = \frac{N_0\Omega}{2} e^{-\Omega|\tau|}$$

show that modulation and diffusion also arise for small and large τ. Show that, for a large τ, diffusion is universal in the sense that it arises for arbitrary modulating noise, not only normal.

Solution For the given correlation function, the integral (5.29) results in

$$\Delta^2(\tau) = \frac{N_0}{\Omega}(\Omega\tau - 1 + e^{-\Omega\tau})$$

For small and large τ, this $\Delta^2(\tau)$ turns into quadratic and linear laws like in (5.34). Therefore, modulation and diffusion arise in the same way.

According to (5.27), the $\Delta\Phi(\tau)$ is a superposition of modulating noises, and for a large τ, many independent noise increments accumulate. Therefore, regardless to the initial distribution, the probability of $\Delta\Phi(\tau)$ is Gaussian because of the central limit theorem (see Problem 5.7), and (5.31) is generally valid.

Problem 5.3

Using correlation properties (5.11) through (5.15), deduce the frequency variance (5.44) from (5.43). Also, deduce the last relation in (5.43).

Solution For $z = x + iy$, we have $|z|^2 = x^2 + y^2$ and $\mathrm{Re}(z^2) = x^2 - y^2$. Therefore $2y^2 = 2\,\mathrm{Im}^2(z) = |z|^2 - \mathrm{Re}(z^2)$, and we apply the identity

$$2\,\mathrm{Im}^2(z) = |z|^2 - \mathrm{Re}(z^2) \quad \text{to} \quad z = \frac{w_n'w - w'w_n}{w^2}$$

Here, w_n is the AS of a random noise whereas w is the AS of a deterministic signal. Besides, according to (5.9) and (5.10), the mean products $\overline{w_n w_n}$, $\overline{w_n' w_n}$, and $\overline{w_n' w_n'}$ are zero, so that $\overline{z^2} = 0$ and $\mathrm{Re}(\overline{z^2}) = 0$. Thus, we obtain

$$2\,\overline{\mathrm{Im}^2(z)} = \overline{|z|^2} = \frac{\overline{|w_n'w - w'w_n|^2}}{|w|^4}$$

that has been used in the last relation of (5.43).

We also have

$$\overline{|w_n' w - w' w_n|^2} = \overline{|w_n'|^2}\,|w|^2 - 2\,\mathrm{Re}[\,\overline{w_n' w_n^*}\,w w'^*] + \overline{|w_n|^2}\,|w'|^2 \quad (5.51)$$

and from (5.11), (5.14), and (5.15) we find

$$\overline{|w_n|^2} = 2\overline{n^2} \qquad \overline{w_n' w_n^*} = 2i\,\overline{\omega}\,\overline{n^2} \qquad \overline{|w_n'|^2} = 2[\Delta\omega^2 + \overline{\omega}^2]\overline{n^2} \quad (5.52)$$

Here, $\overline{\omega}$ and $\Delta\omega$ are the mean spectral frequency of a noise and its bandwidth as given in (5.12) and (5.13). Substituting (5.51) and (5.52) into (5.43) and setting $w = a(t)e^{i\phi(t)}$, $\omega(t) = \phi'(t)$, we come to (5.44).

Problem 5.4

Because of overlap of measuring intervals, frequency measurements of higher precision are for one sampling period. Therefore, for a wide time range, the sampling period Δt should be varied. Then, however, the central frequency ω_0 should also be varied for Nyquist condition $\Delta t < \pi/\omega_0$ to be met. That complicates measuring procedure because a mixer should be tuned for each Δt.

Show that, for frequency variations of a *narrowband* signal (but not its central frequency), Nyquist condition may be violated, and the sampling period $\Delta t \gg \pi/\omega_0$ is allowable. Therefore, the central frequency should not be fitted when the sampling period is varied.

Solution A sampled signal is as follows:

$$u(n\Delta t) = a(n\Delta t)\cos[\omega_0 n\Delta t + \Phi(n\Delta t)] \qquad n = 1, 2, \ldots$$

If a long sampling period $\Delta t = 2\pi N/\omega_0 + \vartheta$ contains $N \geq 1$ periods, we have

$$\begin{aligned} u(n\Delta t) &= a(n\Delta t)\cos[\omega_0 n\vartheta + \Phi(n\Delta t)] \\ &= a(n\Delta t)\cos[(\omega_0 - 2\pi N/\Delta t)n\Delta t + \Phi(n\Delta t)] \end{aligned}$$

So, for a long Δt, the ADC plays a role of the additional mixer with the reference frequency $2\pi N/\Delta t$. This reduces the central frequency of a signal. The Nyquist condition $0 < \omega_1 \pm \Delta\omega < \pi/\Delta t$ must be met for the output signal at the frequency $\omega_1 = \omega_0 - 2\pi N/\Delta t$. That is the condition of spectral nonoverlapping for a mixer (Section 4.1), and the Nyquist band is the output mixer band.

Problem 5.5

For $t, f = 0, 1, \ldots, N-1$, direct and inverse FFTs are given by

$$U(f) = \sum_{t=0}^{N-1} u(t) e^{-\frac{2\pi}{N} ift} \qquad u(t) = \frac{1}{N} \sum_{f=0}^{N-1} U(f) e^{\frac{2\pi}{N} ift} \tag{5.53}$$

For sampled signals, this effective numerical method (especially for $N = 2^n$) is often used instead of the ordinary *Fourier transform* (FT) used for continuous signals. However, FFT and FT are not identical transformations. Consider distinctions between them significant for frequency measurements.

Solution Direct and inverse FFT are exact reciprocal transformations due to orthogonality on the discrete set of points:

$$\sum_{t=0}^{N-1} e^{-\frac{2\pi}{N} if_1 t} e^{\frac{2\pi}{N} if_2 t} = \begin{cases} 0 & \text{for } f_1 \neq f_2 \\ N & \text{for } f_1 = f_2 \end{cases}$$

Dealing with sums, the FFT employs Fourier series instead of Fourier integrals and makes periodic extension in time and frequency (compare Problem 2.7). Therefore, (5.46) is replaced by a circular convolution as follows:

$$w(t) = \sum_{s=0}^{N-1} u(s) \hat{k}[(t-s)_{\text{mod } N}] = \sum_{s=0}^{N-1} u\left[(t-s)_{\text{mod } N}\right] \hat{k}(s)$$

where $(t-s)_{\text{mod } N}$ denotes an integer $(t-s)$ congruous with respect to N. So, for $t \approx 0$, the filter's response $\hat{k}(s)$ overlaps a signal not only around $s = 0$ but also around $s = N-1$ (and the same for $t \approx N-1$). Therefore, edge distortions arise. The linear frequency drift in Figure 5.6 is distorted around edges because of circular overlapping. In Figures 5.3 through 5.5, a part of duration was used (of about 70%), and the distortions are avoided.

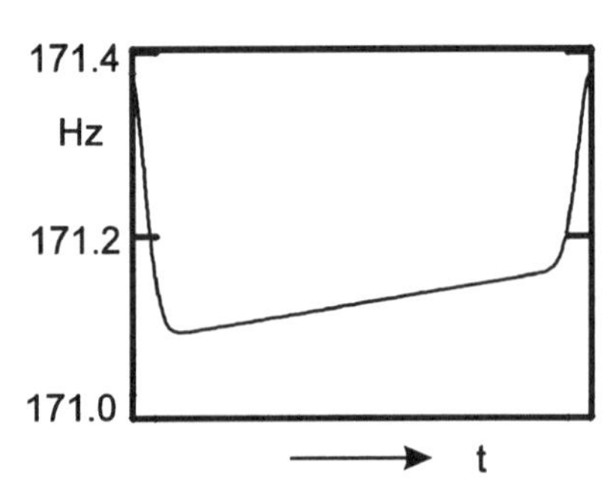

Figure 5.6 Edge frequency distortions for the rate 10^{-8} kHz/ms.

Another distortion arises in spectra. Fourier spectrum of a harmonic signal $e^{2\pi ift}$ consists of one frequency f, but for the FFT, this is true for an integer f only. Otherwise, the samples at $t = 0$ and $t = N - 1$ are unequal, and circular overlapping distorts the spectrum. In Figures 5.3 and 5.4, the actual frequency drift is so small that distortions of the Fourier spectrum should be invisible. Nevertheless, we see distortions of the FFT spectrum. When the direct and inverse FFT are used in pairs, spectral distortions are canceled, and coming back into a time domain, we restore a signal *exactly*. This allows us to measure frequency fluctuations, which are much less than spectral distortions.

Problem 5.6

The idea of global measurements is as follows. We measure each instantaneous frequency during a *long* time T but find the *short-term* frequency change from two frequencies separated by a short time $\Delta T \ll T$. Computer realization provides the highest precision, but the following hardware realization is also informative.

As customary, we use time-counters for N periods and find the frequency from (5.37). However, we use two counters whose long measuring intervals are shifted by $\Delta N \leq N$ periods. Again, we examine a signal with a linear frequency drift in noisy background and find the frequency changes $\delta f(\Delta N)$ for ΔN periods, though each frequency is found for N periods. In the special case $N = \Delta N$, we arrive at local measurements. Explain simulation results given in Figure 5.7 for this hardware method.

Solution In Figure 5.7, standard deviations $\sigma(\Delta N)$ characterize local measurements with $N = \Delta N$. Like in Figure 5.2, it increases as ΔN^{-1}. Moreover, contrary to the actual linear drift, the measured mean frequency change $\overline{\delta f}(\Delta N)$ increases at small ΔN because of noise distortions. However, the curve $\overline{\delta f}(1000)$

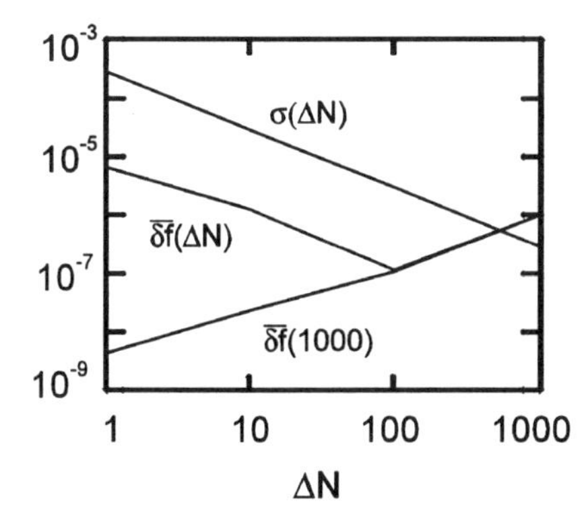

Figure 5.7 Simulation of hardware measurements with two time-counters.

also shows the mean frequency changes for ΔN periods, but each frequency is measured during 1000 periods (global method). Then the curve agrees with the actual drift.

This hardware method is applicable to a linear chirp, but other frequency variations are distorted when averaging in a counter. The AS defines frequency without averaging, and computer measurements are applicable to arbitrary signals.

Problem 5.7

This problem provides definitions related to real random processes. The definitions are not of the most generality but as needed for our reasoning.

Random Processes A random process $u(t)$ is an ensemble of its realizations (sampled functions), which we also denote as $u(t)$. For a fixed instant t_1, the value $u(t_1)$ of realization is random and characterized by the one-dimensional probability density $p(u(t_1))$. The $p(u(t_1))\,du$ is the probability that $u(t_1)$ is between u and $u + du$. Similarly, two random values $u(t_1)$ and $u(t_2)$ are characterized by the two-dimensional probability density $p(u(t_1), u(t_2))$. In general, multidimensional densities of higher orders may also be used, but we use one-dimensional and two-dimensional densities only.

Stationary Processes The process is stationary if its probability densities are invariant to time shifting, so that $p(u(t_1)) = p(u(t_1 - t_0))$ and $p(u(t_1), u(t_2)) = p(u(t_1 - t_0), u(t_2 - t_0))$ for any t_0. This also means that the one-dimensional density $p(u(t))$ is time-independent, and the two-dimensional density $p(u(t_1), u(t_2))$ depends on the difference $t_2 - t_1$ only.

Mean Values (Expectations) The mean value of a one-dimensional function $f(u)$ or a two-dimensional function $g(u_1, u_2)$ is given by

$$\overline{f} = \int_{-\infty}^{\infty} f(u)p(u)\,du \quad \text{or} \quad \overline{g} = \int_{-\infty}^{\infty}\int_{-\infty}^{\infty} g(u_1, u_2)p(u_1, u_2)\,du_1\,du_2$$

Among one-dimensional mean values, the most important are the mean value of a process $\overline{u}$, the mean square of a process $\overline{u^2}$, and the variance σ^2 defined by

$$\overline{u} = \int_{-\infty}^{\infty} up(u)\,du \qquad \overline{u^2} = \int_{-\infty}^{\infty} u^2 p(u)\,du$$

$$\sigma^2 = \int_{-\infty}^{\infty} [u - \overline{u}]^2 p(u)\,du = \overline{u^2} - (\overline{u})^2$$

In general, they depend on the instant t, but for stationary processes, they are time-independent.

Among two-dimensional mean values, the most important is the correlation function:

$$R = \overline{u(t_1)u(t_2)} = \int_{-\infty}^{\infty}\int_{-\infty}^{\infty} u_1 u_2 p(u_1, u_2)\, du_1\, du_2$$

In general, $R = R(t_1, t_2)$ depends on two instants, but for stationary processes, it depends on the difference $R = R(t_2 - t_1)$.

Gaussian (Normal) Processes For a Gaussian process of zero mean, the one-dimensional probability density is as follows:

$$p(u) = \frac{1}{\sqrt{2\pi}\sigma} e^{-u^2/2\sigma^2}$$

In general, the variance σ^2 may be time-dependent. With a proper linear transformation of variables, the multidimensional Gaussian density can be reduced to the product of such one-dimensional densities.

A sum of Gaussian variables $u = u_1 + u_2 + \cdots$ is also Gaussian. Moreover, if independent addends $u_1, u_2, \ldots$ are of arbitrary probability densities but their variances are of the same order of value, the sum of many addends is also Gaussian. This asymptotic statement is known as the *central limit theorem.*

Power Spectral Densities Let us consider a stationary process of zero mean $\overline{u} = 0$. Each realization $u(t)$ of the process is an ordinary deterministic time function, and we can find its spectrum as follows:

$$U(\omega) = \lim_{T\to\infty} \frac{1}{\sqrt{2T}} \int_{-T}^{T} u(t) e^{-i\omega t}\, dt$$

The normalizing factor $1/\sqrt{2T}$ is because the stationary process $u(t)$ does not approach zero for $t \to \pm\infty$, and the integral within infinite limits is divergent (see also Problem 5.1).

For an ensemble of realizations $u(t)$, the spectrum $U(\omega)$ is random at any ω. Its mean value $\overline{U(\omega)}$ is zero because $\overline{u(t)} = 0$. The correlation function depends on the difference $t_2 - t_1$, and we find the power spectral density $S(\omega)$ as follows:

$$\begin{aligned} S(\omega) = \overline{|U(\omega)|^2} &= \lim_{T\to\infty} \frac{1}{2T} \int_{-T}^{T}\int_{-T}^{T} \overline{u(t_1)u(t_2)} e^{-i\omega(t_2 - t_1)}\, dt_1\, dt_2 \\ &= \lim_{T\to\infty} \frac{1}{2T} \int_{-T}^{T}\int_{-T}^{T} R(t_2 - t_1) e^{-i\omega(t_2 - t_1)}\, dt_1\, dt_2 \end{aligned}$$

Now we change the variables t_1, t_2 by the

$$\tau = t_2 - t_1 \quad \text{and} \quad \xi = \frac{t_2 + t_1}{2}$$

Then $dt_1\,dt_2 = d\tau\,d\xi$, and the limits for τ and ξ can be found as follows. For a fixed τ, the coordinate ξ is given by $\xi = t_2 - \tau/2$ or $\xi = t_1 + \tau/2$. From the maximal $t_2 = T$ and the minimal $t_1 = -T$, we then find the upper and lower limits as $\xi = \pm(T - \tau/2)$. On the other hand, the limits $\tau = \pm 2T$ result from the extremal values $t_1 = \pm T$ and $t_2 = \pm T$. Finally, we get

$$\begin{aligned} S(\omega) &= \lim_{T\to\infty} \frac{1}{2T} \int_{-2T}^{2T} R(\tau) e^{-i\omega\tau} d\tau \int_{-(T-\tau/2)}^{(T-\tau/2)} d\xi \\ &= \lim_{T\to\infty} \int_{-2T}^{2T} R(\tau) \left[1 - \frac{\tau}{2T}\right] e^{-i\omega\tau}\, d\tau = \int_{-\infty}^{\infty} R(\tau) e^{-i\omega\tau}\, d\tau \end{aligned}$$

that is, the FT equivalent to (5.2). So, the power spectral density $S(\omega)$ associated with the correlation function $R(\tau)$ is the mean square of the spectrum $\overline{|U(\omega)|^2}$, but special normalization is needed.

References

[1] Rice, S. O., "Mathematical Analysis of Random Noise," *Bell System Tech. J.*, Vol. 23, 1944, p. 282; *ibid.* Vol. 24, 1945, p. 46.

[2] Dugundji, J., "Envelopes and Preenvelopes of Real Waveforms," *IRE Trans. Inf. Theory*, Vol. 4, 1958, pp. 53–57.

[3] Vakman, D., "Computer Measuring of Frequency Stability and the Analytic Signal," *IEEE Trans. Instrum. Meas.*, Vol. 43, 1994, pp. 668–671.

[4] Baghdady, E. J., R. N. Lincoln, and B. D. Nelin, "Short-Term Frequency Stability: Characterization, Theory, and Measurements," *Proc. IEEE*, Vol. 53, 1965, pp. 704–722.

[5] Barnes, J. A., et al., "Characterization of Frequency Stability," *IEEE Trans. Instrum. Meas.*, Vol. IM-20, 1971, pp. 105–120.

[6] Cutler, L. S., and C. L. Searle, "Some Aspects of the Theory and Measurement of Frequency Fluctuations in Frequency Standards," *Proc. IEEE*, Vol. 54, 1966, pp. 136–154.

[7] Hafner, E., "The Effects of Noise in Oscillators," *Proc. IEEE*, Vol. 54, 1966, pp. 179–198.

[8] Lindsey, W. C., and C. M. Chie, "Theory of Oscillator Instability Based upon Structure Functions," *Proc. IEEE*, Vol. 64, 1976, pp. 1652–1666.

[9] Rutman, J., and F. L. Walls, "Characterization of Frequency Stability in Precision Frequency Sources," *Proc. IEEE*, Vol. 79, 1991, pp. 952–960.

[10] Barber, R. E., "Short-Term Frequency Stability of Precision Oscillators and Frequency Generators," *Bell System Tech. J.*, Vol. 50, 1971, pp. 881–915.

[11] Lewis, L. L., "An Introduction to Frequency Standards," *Proc. IEEE*, Vol. 79, 1991, pp. 927–935.

[12] Vessot, R. F. C., "Application of Highly Stable Oscillators to Scientific Measurements," *Proc. IEEE*, Vol. 79, 1991, pp. 1040–1053.

[13] Barnes, J. A., "Atomic Timekeeping and the Statistics of Precision Signal Generators," *Proc. IEEE*, Vol. 54, 1966, pp. 207–220.

[14] Barnes, J. A., and D. W. Allan, "A Statistical Model of Flicker Noise," *Proc. IEEE*, Vol. 54, 1966, pp. 176–178.

[15] Allan, D. W., "Should the Classical Variance be Used as a Basic Measure in Standard Metrology?," *IEEE Trans. Instrum. Meas.*, Vol. IM-36, 1987, pp. 646–654.

[16] Wan, K. W., J. Austin, and E. Vilar, "A Novel Approach to the Simultaneous Measurement of Phase and Amplitude Noise in Oscillators," in *Proc. 44-th Ann. Symp. Frequency Contr.*, 1990, pp. 140–144.

[17] Allan, D. W., "Statistics of Atomic Frequency Standards," *Proc. IEEE*, Vol. 54, 1966, pp. 221–230.

6

Monoharmonic Oscillation Systems

Until now, we have considered signals regardless of the generating systems. Such systems are the subject of the classical oscillation theory originated in the 1930s and then developed by many authors [1–7]. The theory also studies noise effects in oscillators [8–10].

Approximate methods are needed for nonlinear oscillations. Among them, averaging is often used. As early as the 1920s, it was introduced by Van der Pol [11] when he reduced the second-order oscillator equation to the first-order "shortened" equation for amplitude. That was the beginning of oscillation theory. Since then, averaging is often used for defining the APF of nonlinear oscillations.

The AS gives another opportunity. Applying the HT to the initial real equation, we derive an equation for the AS, which defines not only the oscillation itself but also its APF. For slow (low-frequency) modulation, we also apply Bedrosian's theorem (2.22) instead of averaging. Then we derive the known results in a more strict way or obtain new relations of higher precision.

In Section 6.2, we develop oscillator theory and show that amplitude modulation causes frequency fluctuations. This *dynamic* second-order effect is very important for flicker frequency instability. This effect has not been found out with averaging (the first-order method) or *steady-state* methods of higher orders [3,5,7,12]. So the AS noticeably improves oscillator theory.

Higher harmonics arise in nonlinear systems and interact with the main oscillation. Using a special iterative method for *frequency separation* [13], we investigate this interaction in Chapter 7. In this chapter, assuming negligible higher harmonics, we consider monoharmonic oscillating systems. The methods of this chapter are fundamental for more general problems, also.

In Section 6.1 and part of 6.2, we consider *low-frequency* modulation in oscillators. In Section 6.3, the *resonant* exciting frequency is close to oscillator frequency. Finally, in Section 6.4, the excitation is *high-frequency* with respect to oscillations. These frequency distinctions result in dissimilar physical phenomena.

For clarity, we often use pendulums as simple models, but other oscillation systems show similar properties. So, gyro-resonance of electrons displays the same behavior as a nonisochronic pendulum. The generator theory developed in Section 6.2 will be supplemented in Chapter 7.

6.1 Adiabatic Invariant and Momentum

Let us consider the pendulum in Figure 6.1. The length L of its thread varies with a moving carriage, and its linear deviation is $u = L\gamma$, where γ is the angular deviation. For a small γ, the backward force is $-mg\gamma$, and the equation of motion takes the form

$$u''(t) + \Omega^2(t)u(t) = 0 \tag{6.1}$$

where $\Omega(t) = \sqrt{g/L(t)}$ is the nominal pendulum frequency varying with time. For a resonant circuit also shown in Figure 6.1, the charge of a varying capacitor satisfies the same equation.

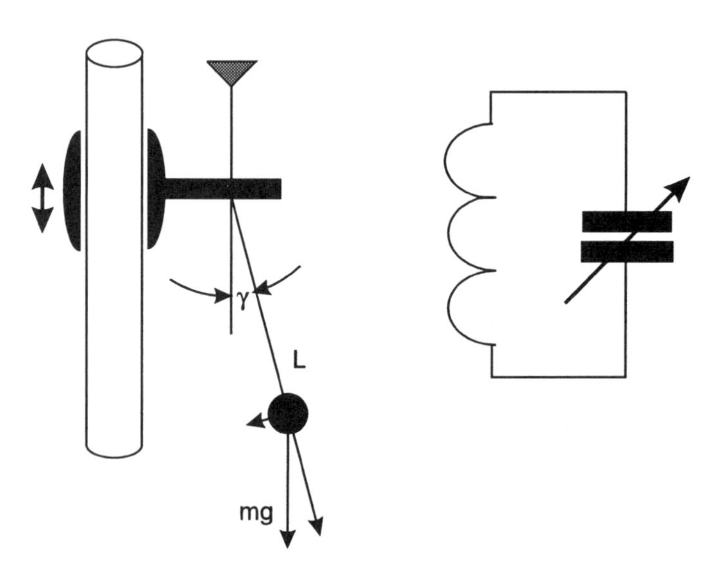

Figure 6.1 Pendulum of a varying length and its electrical analog.

The carriage moving down and up along the rod makes *a* work against the force of thread tension. Therefore, pendulum energy is not conserved. The tension force oscillates as $u^2(t)$. For a quasiharmonic motion with frequency $\Omega(t)$, we can average $u^2(t)$ over a period $2\pi/\Omega$ while the frequency remains constant. Then *adiabatic invariant* is conserved (see Problem 6.1):

$$\Omega(t)a^2(t) = \text{const.} \tag{6.2}$$

Here the term "adiabatic" means that the carriage moves slowly, theoretically, *infinitely* slowly.

The AS provides, however, more general and precise relation. The instantaneous frequency $\omega(t)$ of the pendulum differs from its nominal frequency $\Omega(t)$, and *momentum* is conserved:

$$K = \omega(t)a^2(t) = \text{const.} \tag{6.3}$$

Moreover, slowness is unnecessary, and the modulating carriage frequency may be almost the same as the pendulum frequency. We name K the momentum in the analogy to Kepler's law of areas, where K is the angular momentum of a planet (its sectorial velocity).

6.1.1 Relation to the AS

For this simple system, we derive equation for the AS in detail and show how it works in oscillation problems. First we apply the HT to (6.1) and obtain the equation for the AS $w(t)$ associated with a real motion $u(t)$. Since the HT is commutative with differentiation (property 4 AS, Section 2.2) and if the low-frequency and high-frequency spectra of $\Omega^2(t)$ and $u(t)$ are not overlapping, using Bedrosian's theorem (property 5, Section 2.3), we take the $\Omega^2(t)$ out of the HT and find

$$\mathcal{H}[u'' + \Omega^2 u] = \frac{d^2}{dt^2}\mathcal{H}[u] + \Omega^2\mathcal{H}[u] = v''(t) + \Omega^2(t)v(t) = 0$$

Then, since $w = u + iv$, summing with (6.1), we find the equation for the AS as follows:

$$w''(t) + \Omega^2(t)w(t) = 0 \tag{6.4}$$

Formally, this is the same equation as (6.1), but applied to the complex function $w = ae^{i\phi}$, which defines the amplitude and frequency of oscillations.

Let us multiply (6.4) by the complex conjugate AS $w^*(t)$ and separate real and imaginary parts. Then we come to two equations:

$$\mathrm{Im}(w''w^*) = 0 \quad \text{and} \quad \mathrm{Re}(w''w^*) + \Omega^2|w|^2 = 0 \tag{6.5}$$

Differentiating $w = ae^{i\phi}$, we find $w'w^* = a'a + i\omega a^2$ so that

$$\begin{aligned} K &= \omega a^2 = \mathrm{Im}(w'w^*) \\ \frac{dK}{dt} &= \frac{d}{dt}\,\mathrm{Im}(w'w^*) = \mathrm{Im}(w''w^* + |w'|^2) = \mathrm{Im}(w''w^*) \end{aligned} \tag{6.6}$$

We also find that $\mathrm{Re}(w''w^*) = a''a - \omega^2 a^2$ and the equations (6.5) take the forms

$$\frac{dK}{dt} = 0 \quad \text{and} \quad \omega^2 = \Omega^2 + \frac{a''}{a} \tag{6.7}$$

The first equation (6.7) shows that the conservation law (6.3) is *exact* under spectral nonoverlapping. Therefore, slowness is unnecessary, and modulating frequency of a carriage motion may be lower than, but almost the same as, the pendulum frequency. Clearly, slow adiabatic modulation in (6.2) is a special case.

Using (6.3), we also eliminate the amplitude $a(t)$ from the second equation (6.7) and find

$$\omega^2 = \Omega^2 - \frac{\omega''}{2\omega} + \frac{3}{4}\left(\frac{\omega'}{\omega}\right)^2 \tag{6.8}$$

This equation shows that the actual instantaneous frequency $\omega(t)$ of the pendulum differs from its nominal frequency $\Omega(t)$. For slow modulation, both depend on εt, and the derivatives in the right-hand side are of the order ε^2. Neglecting them, in the *first order* we have $\omega = \Omega$, which results in the adiabatic invariant (6.2). The second-order correction to ω can be found by iteration. Replacing ω by Ω in the small addends of (6.8), we find

$$\omega^2 = \Omega^2 - \frac{\Omega''}{2\Omega} + \frac{3}{4}\left(\frac{\Omega'}{\Omega}\right)^2 \tag{6.9}$$

Continuing the iterations, one can find higher-order corrections with ε^4, ε^6, and so forth. Frequency corrections in (6.9) are of *dynamic* nature: they depend on *derivatives* of modulation and display the transient regime in the oscillator when its frequency is varying. Dynamic frequency distortions of such a kind are important for generators (Section 6.2). We also note that the conservation law (6.3) was found by Kulsrud [14] for more restrictive conditions than nonoverlapping spectra.

6.1.2 Relation to Kepler's Law

In coordinates (u, v), a complex w is a vector while a real Ω is a scalar. Also, ω is the angular velocity of a vector w, and the momentum $K = \omega a^2$ is its sectorial velocity. Therefore, (6.4), when written in the form $w'' = -\Omega^2 w$, defines a motion under the *central force* $-\Omega^2 w$ along the radius to the point w. For a central force, the momentum is conserved, which is the second Kepler's law (law of areas) for planet motion. In fact, for two dimensions, we have derived this law from (6.4).

However, the Kepler's law is true for the *complex* equation (6.4) that results from the real equation (6.1) for low-frequency modulation and nonoverlapping spectra. We will show in Chapter 7 that the same real equation results in a very different complex equation for high-frequency modulation. Then parametric resonances arise in the same physical system.

6.2 Narrowband Generator Theory

We now address the generator in Figure 6.2. In this section, we assume that its *narrowband* amplifier (a current source) does not permit higher harmonics to pass. Then the output current is $I(t) = k(a)u(t)$, where u is the input voltage and $k(a)$ is the gain-factor dependent on the input amplitude. We also assume that, because of saturation, the output amplitude is constant and the gain-factor is $k = 1/a$ (Figure 6.2). Generalizing, we set $k = m(\mu t)/a$, where $m(\mu t)$ is additional slow modulation with $\mu \ll 1$.

Slow modulation $m(\mu t)$ results in amplitude variations, but according to the classical first-order theory, frequency of this *isochronic* generator remains constant [1,4,8–11]. We will show, however, that amplitude modulation is accompanied with additional frequency variations [15]. This *dynamic* second-order effect comes from a transient process associated with modulation. Though very

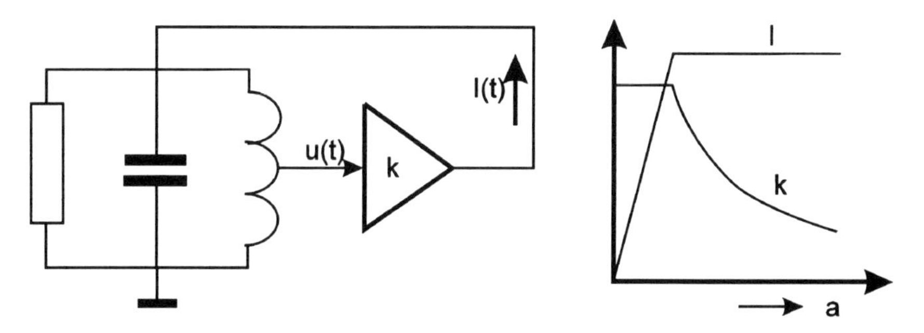

Figure 6.2 Generator circuit.

important for frequency stability, it has not been found with averaging (the first-order method) or steady-state methods of higher orders [3,7,12].

6.2.1 Dynamic Frequency Distortions

According to Figure 6.2, $I = \frac{m}{a}u = I_L + I_C + I_R$, where the currents in the inductor, capacitor, and resistor are related as

$$u = L\frac{dI_L}{dt} \qquad I_C = C\frac{du}{dt} \qquad u = RI_R$$

Therefore, the generator obeys the following equation:

$$u'' + u = \varepsilon\frac{d}{dt}(\gamma u), \quad \text{where } \varepsilon = Q^{-1} \ll 1 \quad \text{and} \quad \gamma = \frac{m(\mu t)}{a} - 1 \qquad (6.10)$$

Here, we introduce such a timescale that the resonant frequency $\omega_0 = 1/\sqrt{LC}$ becomes a unit. Quality Q of a resonant circuit is large, and $\varepsilon \ll 1$ shows that its bandwidth is small with respect to ω_0. The function γ describes saturation nonlinearity, additional slow modulation, and damping in the circuit. We emphasize, however, that our main result is independent of a particular function γ.

In the same way as for the pendulum in Section 6.1, we now apply the HT and Bedrosian's theorem to (6.10). Since the low-frequency and high-frequency spectra of γ and u are nonoverlapping, we take γ out of the HT and come to an equation for the AS in the form

$$w'' + w = \varepsilon\frac{d}{dt}(\gamma w) = \varepsilon(\gamma' w + \gamma w') \qquad (6.11)$$

Then, again as before, we multiply this equation by w^* and separate the real and imaginary parts. According to (6.6), $\mathrm{Im}(w'w^*) = K = \omega a^2$ and $\mathrm{Im}(w''w^*) = K'$, and we obtain two equations:

$$\begin{aligned} K' &= \varepsilon\gamma K \\ \omega^2 &= 1 - \varepsilon\gamma' - \varepsilon\gamma\frac{a'}{a} + \frac{a''}{a} \end{aligned} \qquad (6.12)$$

Now the momentum K is not conserved because of generator feedback.

Frequency deviations are small, and if we set $\omega = 1$ and $K = a^2$, the first equation (6.12) turns into Van der Pol's equation for amplitude:

$$\frac{da^2}{dt} = \varepsilon\gamma a^2 \quad \text{or} \quad \frac{da}{dt} = \frac{\varepsilon}{2}\gamma a$$

In the first order, this equation is usually derived with averaging, and slowness is assumed. However, equations (6.12) are exact for nonoverlapping spectra.

Frequency corrections in the second equation (6.12) are of the order of ε^2 (since a and γ depend on εt). Therefore, to define the frequency, we express $\varepsilon\gamma$ from the first equation (6.12) and replace K by a^2 so that

$$\varepsilon\gamma = \frac{K'}{K} = 2\frac{a'}{a} \qquad \varepsilon\gamma' = 2\left[\frac{a''}{a} - \left(\frac{a'}{a}\right)^2\right]$$

Then, the second equation (6.12) takes a simple form:

$$\omega^2 = 1 - \frac{a''}{a} \tag{6.13}$$

which shows a new important effect: *amplitude variations cause frequency deviations.*

Interpretation Let us investigate the converse: the amplitude is varying while the frequency remains constant, $w(t) = a(t)e^{it}$. Then the phase and frequency of the derivative $w'(t)$ are also varying:

$$w' = i(a - ia')e^{it}$$

$$\omega = 1 - \frac{d}{dt}\arctan\frac{a'}{a} \approx 1 - \frac{a''}{a} + \left(\frac{a'}{a}\right)^2$$

and analogous variations appear in the second derivative. As a result, the AS $w(t)$ of a constant frequency cannot satisfy (6.11).

This can be shown as follows. Substituting $w(t) = a(t)e^{it}$ and its derivatives into (6.11), we separate real and imaginary parts and come to two equations:

$$a'' = \varepsilon\frac{d}{dt}(\gamma a) \quad \text{and} \quad a' = \frac{\varepsilon}{2}\gamma a$$

The second one is Van der Pol's equation for amplitude. Differentiating it, we find $a'' = \frac{\varepsilon}{2}\frac{d}{dt}(\gamma a)$, which is incompatible with the first equation, excepting $\gamma a = \text{const}$ and $a'' = 0$. Then, however, $\omega = 1$ according to (6.13).

So, slow amplitude modulation causes small phase differences between oscillation and its derivatives, and dynamic frequency fluctuations (6.13) appear to support the *phase balance* in the oscillator. Relation (6.13) between amplitude and frequency is independent of particular modulation or nonlinearity. To conserve the phase balance, any amplitude variation (more precisely, its second derivative) causes frequency distortion. Because, however, frequency corrections are of the second order, within the first-order classical theory, the generator is isochronic.

6.2.2 Flicker-Noise Effect

Slow flicker noise $n(\mu t)$ of the spectral density

$$N(\omega) = \frac{B}{\omega} \quad \text{where we set } 0 < \omega < 1 \tag{6.14}$$

commonly exists in a supply voltage and modulates the amplifier gain. This *quasistatic* modulation is slow with respect to a transient time (of about $1/\varepsilon$) of a generator, and $\mu \ll \varepsilon$ in (6.10). Nevertheless, we restrict the noise band to high (unit) generator frequency $\omega_0 = 1$, though it is often restricted to a much lower frequency. The point is that derivatives of the noise are significant for frequency fluctuations, and high-frequency noise components should be taken into consideration. In Chapter 7 (Problem 7.3), it is also shown that modulating frequencies higher than $\omega = 1$ should not be considered.

We also estimate $B \sim 10^{-14}$ in (6.14) that comes from the following. Since $\overline{n^2} = \frac{1}{\pi}\int N(\omega)\, d\omega$, dimension of the spectral density $N(\omega)$ is V^2/Hz and dimension of B is V^2. In the natural scale, the value $B \sim 10^{-12} - 10^{-14}\ V^2$ has been established experimentally. We use normalized equations with the generator frequency $\omega_0 = 1$ and its amplitude $a = 1$. Therefore, we have to replace the natural B by the B/a^2 where a is the natural amplitude in volts. For $a \sim 1{-}10$ V, this results in $B \sim 10^{-14}$.

So, in (6.10) we set $m = 1 + n(\mu t)$, $\gamma = \frac{1+n(\mu t)}{a} - 1$, and because $K = \omega a^2$ and $a = \sqrt{K/\omega}$, the first equation (6.12) takes the form

$$K' = \varepsilon \left(\frac{1 + n(\mu t)}{\sqrt{K}} \sqrt{\omega} - 1 \right) K \tag{6.15}$$

With an error of ε^3, we set $\omega = 1$, and (6.15) turns into standard Bernoulli equation (see Problem 6.7). Substituting a^2 for K, we reduce it to the linear equation

$$a' + \frac{\varepsilon}{2} a = \frac{\varepsilon}{2}[1 + n(\mu t)]$$

with the general solution as follows:

$$a = 1 + \frac{\varepsilon}{2} e^{-\varepsilon t/2} \int n(\mu t) e^{\varepsilon t/2}\, dt + C e^{-\varepsilon t/2}$$

Here, a transient process (the last addend) can be neglected after a while (see Problem 7.5 for transient regimes after switching on). Then, integrating by parts, we obtain

$$a = 1 + n(\mu t) - 2\frac{dn(\mu t)}{d(\varepsilon t)} + 4\frac{d^2 n(\mu t)}{d(\varepsilon t)^2} + \cdots$$

where derivatives are negligible for quasistatic modulation with $\mu \ll \varepsilon$.

Finally, the amplitude and frequency are given by

$$a = 1 + n(\mu t) \qquad \omega = 1 - \frac{n''(\mu t)}{2} \tag{6.16}$$

where the frequency is found from (6.13) for $n \ll 1$. As expected, amplitude variations follow slow noise modulation. However, dynamic frequency variations follow the *second derivative* of the noise and result in unusual effects.

Fast Fluctuations From Slow Modulation The amazing fact is that slow flicker noise causes fast frequency fluctuations. Indeed, a narrowband amplitude spectrum is the same as for the modulating flicker noise $n(t)$, but frequency fluctuations follow $n''(t)$, and their spectral density is proportional to ω^4. Therefore, from (6.14) and (6.16), we find

$$S_a(\omega) = \frac{B}{\omega} \qquad S_\omega(\omega) = \frac{\omega^4 S_a(\omega)}{4} = \frac{B\omega^3}{4} \qquad S_\phi(\omega) = \frac{S_\omega(\omega)}{\omega^2} = \frac{B\omega}{4} \tag{6.17}$$

So, the spectral densities for frequency and phase are *increasing* throughout the band $0 < \omega < 1$, and high-frequency fluctuations dominate.

The phase spectral density $S_\phi(\omega)$ differs from that assumed in instability theory (compare Section 5.3.2). Besides, S_ω is not singular, and the standard frequency variance is obtainable instead of Allan's variance (5.40). So we have

$$\sigma_\omega^2 = \frac{1}{\pi}\int_0^1 S_\omega(\omega)\, d\omega = \frac{B}{4\pi}\int_0^1 \omega^3 d\omega = \frac{B}{16\pi} \tag{6.18}$$

For $B \sim 10^{-14}$, we estimate $\sigma_\omega \sim 10^{-8}$ that is typical for good oscillators.

Flicker frequency instability is a modulation effect independent of a circuit quality Q. Therefore, contrary to commonly held views [16], we cannot reduce it with a high-quality resonator. However, frequency fluctuations caused by slow amplitude variations can be reduced with a servo-system for amplitude stabilization. Such servo-systems are frequently employed in precision oscillators, and we now understand how they work.

Spectral Line Shape Flicker noise is often assumed to be Gaussian, and we find the generator spectral line using Section 5.2. For the frequency density (6.17), the mean-square phase increment (5.30) is as follows:

$$\Delta^2(\tau) = \frac{B}{\pi}\int_0^1 \omega \sin^2\frac{\omega\tau}{2}\,d\omega = 4\sigma_\omega^2\left(1 - \frac{2}{\tau}\sin\tau + \frac{4}{\tau^2}\sin^2\frac{\tau}{2}\right)$$

and the spectrum (5.32) takes the form

$$\begin{aligned} S_z(\omega) &= \frac{1}{2}\int_0^\infty e^{-\Delta^2(\tau)/2}\cos\omega\tau\,d\tau \\ &\approx \frac{1}{2}\int_0^\infty \left[1 - 2\sigma_\omega^2\left(1 - \frac{2}{\tau}\sin\tau + \frac{4}{\tau^2}\sin^2\frac{\tau}{2}\right)\right]\cos\omega\tau\,d\tau \\ &= \pi\left(\frac{1-2\sigma_\omega^2}{2}\delta(\omega) + \sigma_\omega^2|\omega|\right) \quad \text{for } -1 < \omega < 1 \end{aligned} \tag{6.19}$$

where we used the first-order expansion of the exponential function.

The spectrum (6.19) consists of the δ function and small wings increasing over the whole modulation band $-1 < \omega < 1$ (see Figure 6.3). The wings result from the rectangular spectrum of $\frac{2}{\tau}\sin\tau$ and the triangular spectrum of $\frac{4}{\tau^2}\sin^2\frac{\tau}{2}$ of the same unit band. Their difference gives the wings $|\omega|$. Because of amplitude modulation in (6.16), the spectrum (6.19) is convolved with a narrowband amplitude spectrum. Then the δ function turns into the flicker spectrum (6.14), but the wideband wings remain the same. Note also that the spectra of amplitude and frequency fluctuations are almost nonoverlapping and, therefore, uncorrelated.

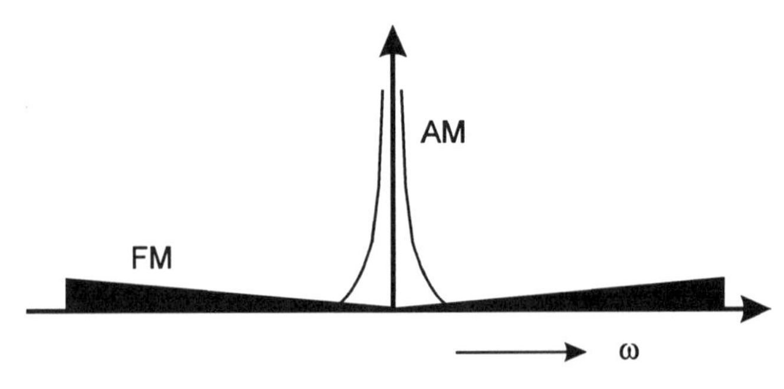

Figure 6.3 Flicker spectral line.

The resulting spectrum in Figure 6.3 explains another method for frequency stabilization. An external filter removes the wings and deletes frequency fluctuations. The filtering is effective because frequency modulation is wideband whereas amplitude modulation is narrowband. Such filters are also used in precision oscillators. It may be remarked that the Lorentzian spectral line is often assumed for generators [8,10]. As shown below, however, this is true for the additive but not modulating noise.

6.2.3 Wideband Additive Noise Effect

If a weak wideband noise $n(t)$ is added to the amplifier current, so that $I = ku(t) + n(t)$ for $k = 1/a$, (6.11) takes the form

$$w'' + \varepsilon w' + w = \varepsilon \frac{d}{dt}\left(\frac{w}{a}\right) + \varepsilon w_n' \tag{6.20}$$

Here, w_n is the AS related to a noise n, and we have moved the dissipative term $\varepsilon w'$ into the left-hand side of equation.

The two addends in the right-hand side belong to separate frequency bands. In fact, the $w(t)$ is a narrowband resonant signal around $\omega_0 = 1$, whereas the noise $w_n(t)$ is wideband. Their spectra are nonoverlapping or, at least, the noise is mostly out of a resonant band. Therefore, w in the left-hand side also consists of two components, $w = w_r + w_o$, where subscripts denote the resonant and outside bands. Substituting this sum into the left-hand side of (6.20) and separating the components of different frequencies, we find two equations:

$$w_r'' + \varepsilon w_r' + w_r = \varepsilon \frac{d}{dt}\left(\frac{w_r}{a}\right) \tag{6.21}$$

$$w_o'' + \varepsilon w_o' + w_o = \varepsilon w_n' \tag{6.22}$$

This *frequency separation procedure* will be studied in Chapter 7. Now we complete the solution. Equation (6.21) is satisfied with the harmonic AS $w_r = e^{it}$ of unit amplitude and unit frequency. On the other hand, (6.22) describes filtering of the input noise $\varepsilon w_n'$ in the resonant circuit of a generator. The transfer function of the circuit

$$K(\omega) = \frac{1}{1 - \omega^2 + i\varepsilon\omega} \approx \frac{1}{2\omega(1 - \omega) + i\varepsilon\omega}$$

is concentrated around $\omega = 1$, and the spectral density of w_n' is $\omega^2 N$, where N is the constant density of w_n (at $\omega > 0$). Therefore, the output Lorentzian spectrum is as follows:

$$S_o(\omega) = \omega^2 \varepsilon^2 N |K(\omega)|^2 = \frac{\varepsilon^2 N}{4(1-\omega)^2 + \varepsilon^2} \tag{6.23}$$

Finally, the resulting AS is the sum of the resonant signal and filtered noise:

$$w = e^{it} + w_o \tag{6.24}$$

So, generator nonlinearity defines the unit amplitude of the resonant signal but does not influence upon the output noise that is the same as for the linear resonant circuit.

Noise distortions are of the first order, and some results obtained with averaging [8–10] are similar to ours. Averaging is effective for slow modulation, however, and a narrowband *resonant* noise has only been considered. We will show, however, that the noise components *outside* a resonant band are more important for frequency fluctuations.

Amplitude and Frequency Distortions The resulting AS (6.24) can be written as

$$w = e^{it} + (x + iy)e^{it} = (1 + x + iy)e^{it}$$

and its APFs are as follows:

$$a = \sqrt{(1+x)^2 + y^2} \approx 1 + x \quad \Phi = \arctan\frac{y}{1+x} \approx y \quad \delta\omega = \Phi' \approx y' \tag{6.25}$$

Here, x and y are small noise quadratures, and their spectral densities are given by (5.23). Using (6.23) as the S_w in (5.23), we find

$$S_a = S_\Phi = \frac{\varepsilon^2 N}{4\omega^2 + \varepsilon^2} \tag{6.26}$$

and therefore

$$S_\omega = \omega^2 S_\Phi = \frac{\omega^2 \varepsilon^2 N}{4\omega^2 + \varepsilon^2} \tag{6.27}$$

Lorentzian spectral densities (6.26) for amplitude and phase are *narrowband* with $\Delta\omega \sim \varepsilon$. However, the spectral density S_ω timed by ω^2 is *wideband*, and nonresonant noise components are significant for frequency fluctuations.

Restricting the input noise to the band $0 < \omega < 2$ (double generator frequency, see Problem 7.4), we obtain amplitude and frequency variances as follows:

$$\sigma_a^2 = \frac{1}{\pi}\int_0^1 \frac{\varepsilon^2 N}{4\omega^2 + \varepsilon^2}\, d\omega = \frac{\varepsilon N}{4} \tag{6.28}$$

$$\sigma_\omega^2 = \frac{1}{\pi}\int_0^1 \frac{\omega^2\varepsilon^2 N}{4\omega^2 + \varepsilon^2}\, d\omega = \frac{\varepsilon^2 N}{4\pi} \tag{6.29}$$

For typical values $N \sim 10^{-12}$ (see Problem 6.5) and $\varepsilon \sim 10^{-3}$, we have $\sigma_\omega \sim 10^{-10}$, which is much less than for flicker fluctuations. Also, σ_ω depends on ε, and for precision generators with $\varepsilon \sim 10^{-6}$, additive frequency distortions are negligible.

Resume Frequency noise distortions have been studied since the 1960s. The first-order additive noise cannot explain actual instability of generators, however, and its effective band has not been considered properly. On the other hand, second-order flicker frequency distortions could not be studied with known methods. We found out that flicker frequency distortions dominate. For precision generators, they agree with actual instability and reassert practical methods for frequency stabilization. The most noticeable point is that slow flicker noise may result in fast frequency fluctuations.

6.3 Resonant Excitation of Nonlinear Oscillators

We now address very different oscillators characterized with two features:

- They are *nonisochronic*, and their frequency is amplitude-dependent.
- The exciting external force is of a resonant frequency.

When amplitude increases, frequency deviates from a resonance. Therefore, two oscillations are superposed: of a resonant exciting frequency and of a varying oscillator frequency. Their nonlinear interaction displays interesting and surprising behavior.

6.3.1 Nonisochronic Pendulum

The pendulum under consideration obeys the equation:

$$u'' + \Omega^2(a)u = \varepsilon F \cos t \qquad \varepsilon \ll 1 \qquad \Omega^2 = 1 - \varepsilon a^2 \tag{6.30}$$

Here, εF is a small exciting force at $\omega = 1$, but the oscillator frequency Ω is amplitude-dependent. The dependence in (6.30) is typical for nonlinear oscillators (see Chapter 7).

Like in Section 6.1, Ω is a slow function, and spectra of Ω^2 and u are nonoverlapping. Therefore, applying the HT, we come to an equation for the AS:

$$w'' + (1 - \varepsilon a^2)w = \varepsilon F e^{it} \tag{6.31}$$

Coming to the AS, we also replaced $\cos t$ by e^{it}. Then, setting $w = ze^{it}$ and differentiating, we have $w' = (z' + iz)e^{it}$, $w'' = (z'' + 2iz' - z)e^{it}$, and the equation for the complex envelope $z = x + iy$ dependent on εt takes the form

$$\varepsilon \frac{d^2 z}{d(\varepsilon t)^2} + 2i \frac{dz}{d(\varepsilon t)} = a^2 z + F$$

In the first order, we also neglect the second derivative with a small factor ε and, separating the real and imaginary parts, we obtain the system for quadratures as follows:

$$\frac{dx}{d(\varepsilon t)} = \frac{x^2 + y^2}{2} y \qquad -\frac{dy}{d(\varepsilon t)} = \frac{F}{2} + \frac{x^2 + y^2}{2} x \tag{6.32}$$

This system has a rest point at $x_* = -F^{1/3}$ and $y_* = 0$, where derivatives are zero. Therefore, harmonic oscillation of the constant amplitude $a_* = F^{1/3}$ is admissible, but it appears for special initial conditions. In general, the system (6.32) can be solved as follows.

Let us introduce the function

$$D(x, y) = \frac{F}{2} x + \frac{(x^2 + y^2)^2}{8}$$

and consider its partial derivatives

$$\frac{\partial D}{\partial x} = \frac{F}{2} + \frac{x^2 + y^2}{2} x \qquad \frac{\partial D}{\partial y} = \frac{x^2 + y^2}{2} y$$

Comparing with (6.32), we see that

$$\frac{dx}{d(\varepsilon t)} = \frac{\partial D}{\partial y} \qquad \frac{dy}{d(\varepsilon t)} = -\frac{\partial D}{\partial x}$$

$$\frac{dD}{d(\varepsilon t)} = \frac{\partial D}{\partial x} \cdot \frac{dx}{d(\varepsilon t)} + \frac{\partial D}{\partial y} \cdot \frac{dy}{d(\varepsilon t)} = 0$$

Therefore, the trajectory of motion is the level curve $D(x, y) = \text{const}$. For zero initial conditions $x(0) = y(0) = 0$, the curve $D(x, y) = 0$ obeys the equation

$$x = -\frac{(x^2 + y^2)^2}{4F} \quad \text{or} \quad \cos\Phi = -\frac{a^3}{4F} \tag{6.33}$$

This is shown in Figure 6.4.

The pendulum motion can be explained as follows. For the linear pendulum with $\Omega = 1$, (6.31) has the solution

$$w(t) = -\frac{i\varepsilon Ft}{2}e^{it}$$

This oscillation is shifted by $-\pi/2$ with respect to the force F, and its amplitude increases linearly. For small amplitudes, nonlinearity is small, and swinging is the same as for a linear oscillator. Really, in Figure 6.4, the trajectory is shifted by $-\pi/2$ at the origin. When increasing amplitude, oscillator frequency deviates from the resonance, and the amplitude grows slower. Then a representative point

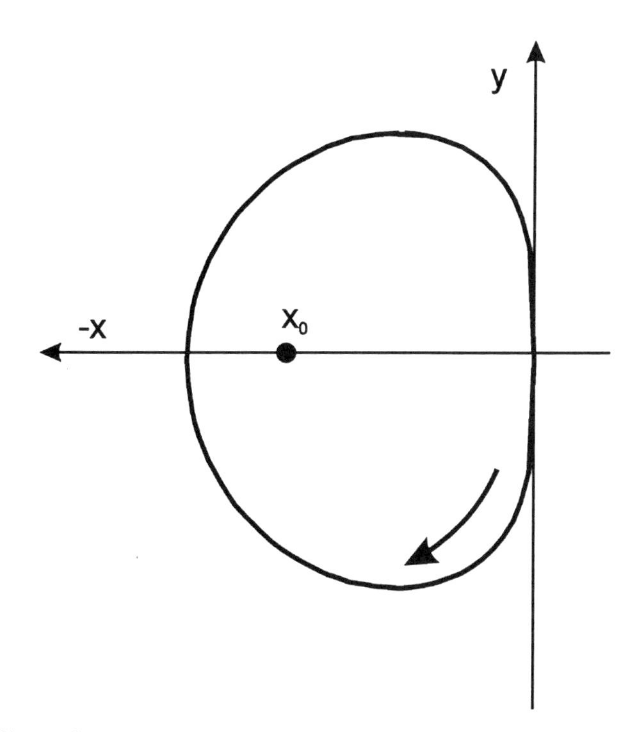

Figure 6.4 Slow trajectory.

moves along the lower side of the trajectory, and the phase varies from $-\pi/2$ to $-\pi$. When it achieves $-\pi$, the force F becomes braking, and amplitude decreases. While a representative point moves along the upper side of the trajectory, the phase varies from $-\pi$ to $-3\pi/2$, and the amplitude decreases from maximum to zero. At zero amplitude, the phase jumps from $-3\pi/2$ to $-\pi/2$, and the cycle starts again. For a dissipative pendulum, the trajectory becomes spiral and approaches the point x_* of a sinusoidal motion.

6.3.2 Gyro-Resonance of Electrons

Lorentzian force acting on the electron is $\mathbf{F} = e\mathbf{E} + \frac{e}{c}\mathbf{v} \times \mathbf{B}$, where e is the electron charge, $\mathbf{E}$ and $\mathbf{B}$ are the vectors of electric and magnetic fields, $\mathbf{v}$ is an electron velocity (also a vector), and the sign $\times$ denotes the vector product. The two terms define electric and magnetic forces. The magnetic force is orthogonal to the velocity $\mathbf{v}$ and proportional to it.

For plane electric field $\mathbf{E}(u, v)$ (u and v are rectangular coordinates) and for the magnetic field $\mathbf{B}$ perpendicular to the plane, complex notations can be used. Then the equation of motion $m\mathbf{w}'' = \mathbf{F}$, where m is mass and $\mathbf{w}$ is the coordinate of the electron, can be written as

$$w'' = \frac{e}{m}E - i\omega_c w' \qquad \omega_c = \frac{eB}{mc} \tag{6.34}$$

Here, $w = u + iv$ is a complex coordinate of the electron (not AS), $E = E_u + iE_v$ is a complex electric field, and ω_c is the cyclotron frequency. The $-i\omega_c w'$ is the velocity-dependent magnetic force, and its orthogonality to the velocity w' is now shown by the factor i.

The mass m of a *relativistic* electron is velocity-dependent, so that

$$\omega_c = \frac{eB}{m_0 c\sqrt{1 + |w'|^2/c^2}} \approx \omega_0 \left(1 - \frac{|w'|^2}{2c^2}\right)$$

where m_0 is the rest mass and c is the light speed. We also assume the rotating electric vector $E = E_0 e^{-i\omega_0 t}$ of a resonant frequency ω_0. Then, using dimensionless time $\omega_0 t$, we rewrite (6.34) in the form

$$w'' + i\left(1 - \frac{\omega_0^2}{2c^2}|w'|^2\right) w' = Fe^{-it} \quad \text{where } F = \frac{eE_0}{m\omega_0^2} = \frac{mc^2 E_0}{eB^2} \tag{6.35}$$

is the amplitude of the exciting electric force.

Finally, we introduce a *slow velocity* z added to the basic rotation and its quadratures by setting $w' = ze^{-it}$, $z = x + iy$. Then, substitution into (6.35) results in equations for quadratures:

$$x' = F - \frac{\omega_0^2}{2c^2}(x^2 + y^2)y \qquad y' = \frac{\omega_0^2}{2c^2}(x^2 + y^2)x \tag{6.36}$$

These equations are similar to (6.32), and therefore, gyro-resonance is a nonlinear interaction of the exciting rotating field and circular electron motion with a varying frequency (due to relativistic mass variations). Like the amplitude of the pendulum considered, the electron orbit enlarges and shrinks (down to zero) by turns.

It may be remarked that the AS has not been used here, and electron coordinates $u(t)$ and $v(t)$ are not related with the HT a priori. If orbital variations are slow with respect to fast rotation in a strong magnetic field, however, spectra are nonoverlapping. Then the *solution* $w(t)$ is the AS, and $v(t)$ is the HT of $u(t)$. Now the AS is not a definition of amplitude and frequency but the property of the actual motion.

6.3.3 Pendulum Between Reflecting Walls

The pendulum in Figure 6.5(a) moves between rigid reflecting walls at $u = \pm 1$. For a small initial velocity, the pendulum does not touch the walls, and its sinusoidal motion is of the unit frequency $\omega = 1$. For a large initial velocity,

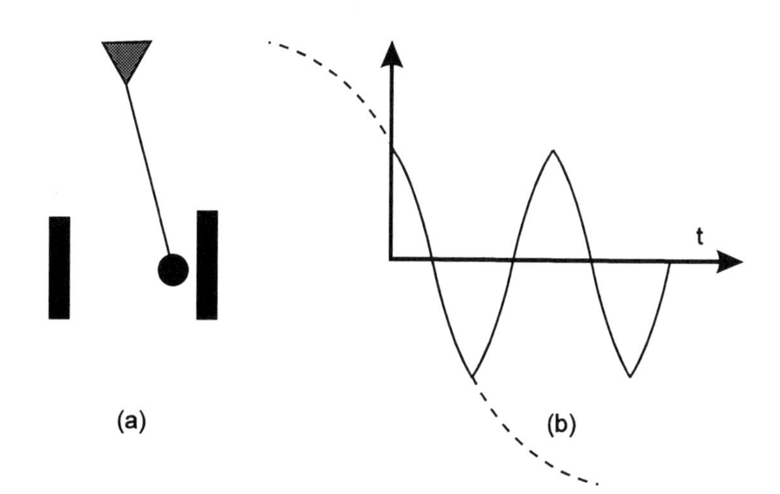

Figure 6.5 Pendulum between reflecting walls.

the pendulum is reflected from walls (elastic blows), and its motion consists of sinusoidal segments shown in Figure 6.5(b).

Taking $t = 0$ at the instant of a blow on the right wall $u = 1$, we have $u = A\sin(\alpha - |t|)$, where $A\sin\alpha = 1$ and $|t| < \alpha$ or, using the cutoff angle $\Delta = \pi/2 - \alpha$, we have

$$u = \frac{\cos(|t| + \Delta)}{\cos\Delta} \quad \text{for } |t| < \pi - 2\Delta \tag{6.37}$$

When blowing, period of oscillations is reduced to $T = 2(\pi - 2\Delta)$, and this periodic motion can be written as a Fourier series:

$$u(t) = a_1 \cos\Omega t + a_3 \cos 3\Omega t + \cdots$$

where, for $k = 1, 3, \ldots$, harmonic amplitudes are given by

$$a_k = \frac{2\Omega}{\pi}\int_0^{\pi/\Omega} \frac{\cos(t + \Delta)}{\cos\Delta} \cos k\Omega t\, dt = \frac{4\tan\Delta}{\pi} \frac{1 - 2\Delta/\pi}{k^2 - (1 - 2\Delta/\pi)^2}$$

For this pendulum, nonlinearity is strong, and its motion greatly differs from a sinusoid. Frequency and amplitude of the *first harmonic*

$$\Omega = \frac{1}{1 - 2\Delta/\pi} \qquad a = \frac{\tan\Delta}{\Delta} \frac{1 - 2\Delta/\pi}{1 - \Delta/\pi} \tag{6.38}$$

are interrelated as shown in Figure 6.6(a). Because of strong nonlinearity, the dependence $\Omega(a)$ consists of two parts. For small deviations, when the pendulum

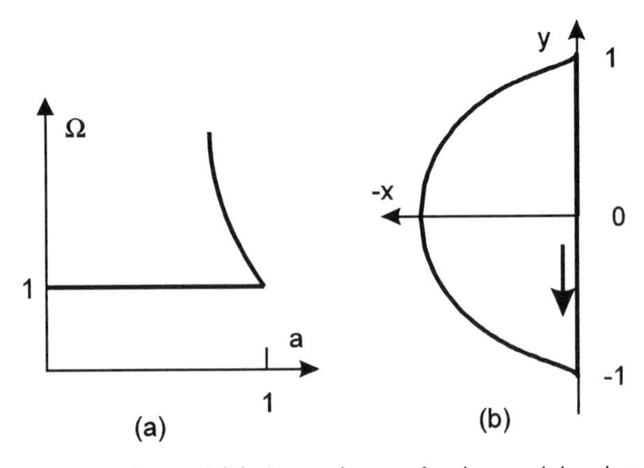

Figure 6.6 (a) Nonisochronism and (b) slow trajectory for the pendulum between reflecting walls.

does not touch the walls, its frequency is constant, but for large deviations, frequency increases while amplitude decreases.

Now, considering excitation with a resonant force, we come to the equation for the AS of the first harmonic analogous to (6.31):

$$w'' + \Omega^2(a)w = Fe^{it} \tag{6.39}$$

and its solution is similar. Using the complex amplitude $z(t) = w(t)e^{-it}$, we have

$$z'' + 2iz' + [\Omega^2(a) - 1]z = F$$

and neglecting the small second derivative, we come to equations for quadratures similar to (6.32):

$$x' = -\frac{1}{2}[\Omega^2(a) - 1]y \qquad -y' = \frac{F}{2} - \frac{1}{2}[\Omega^2(a) - 1]x \tag{6.40}$$

Further, for the function

$$D(x, y) = \frac{F}{2}x + \frac{1}{4}\int_{a^2}^{1}[\Omega^2(a) - 1]d(a^2) \qquad a^2 = x^2 + y^2$$

partial derivatives are as follows:

$$\frac{\partial D}{\partial x} = \frac{F}{2} - \frac{1}{4}[\Omega^2(a) - 1]\frac{\partial a^2}{\partial x} = \frac{F}{2} - \frac{1}{2}[\Omega^2(a) - 1]x = -y'$$

$$\frac{\partial D}{\partial y} = -\frac{1}{4}[\Omega^2(a) - 1]\frac{\partial a^2}{\partial y} = -\frac{1}{2}[\Omega^2(a) - 1]y = x'$$

$$\frac{dD}{dt} = \frac{\partial D}{\partial x}x' + \frac{\partial D}{\partial y}y' = 0$$

The slow trajectory $D(x, y) = 0$ is shown in Figure 6.6(b). While no blows, a representative point moves along the y-axis from $y = 0$ to $y = -1$, and amplitude grows linearly up to the maximum $a = 1$. Then, when blowing, increasing frequency deviates from a resonance, the amplitude (of the first harmonic) decreases, and the phase varies from $-\pi/2$ to $-\pi$. At this point, the cutoff angle Δ is maximal, and the amplitude is minimal. Further, in the upper side of the trajectory, the frequency decreases to $\omega = 1$, the amplitude increases to $a = 1$, and the phase varies from $-\pi$ to $-3\pi/2$. When going along the y-axis from 1 to 0 with no blows again, the frequency is constant, and the amplitude decreases linearly. At zero amplitude, the phase jumps by π as before, and the cycle starts again.

For this pendulum, amplitude and frequency variations are more complicated than in Section 6.3.1. In Figure 6.4, during a semicycle from $-\pi/2$ to $-\pi$, amplitude regularly increases while frequency decreases. Now the semicycle consists of two parts: first amplitude increases up to the maximum at a constant frequency, and then it decreases while frequency increases.

6.4 High-Frequency Excitation

At its low position, the pendulum in Figure 6.7 crosses the field of the electromagnet. The electromagnet is fed with a high-frequency 60 Hz, whereas pendulum frequency is about 1 Hz. Nevertheless, the pendulum swings with a high-frequency force up to considerable amplitude. Moreover, the amplitude is highly stable and remains constant when the electromagnet current varies in a wide range. On the other hand, many stable amplitudes exist, and the pendulum jumps from one to another if the current is varying further [6]. Let us comprehend this unusual behavior of the pendulum.

6.4.1 Mechanism of Excitation

Although the feeding current is sinusoidal and the magnetic field is quasistatic, the moving pendulum modulates the acting force. This force is a train of short

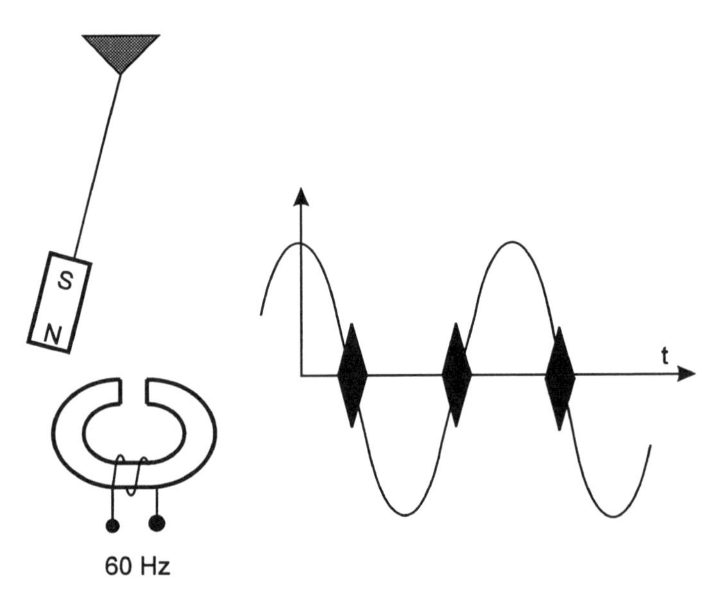

Figure 6.7 Excitation of a pendulum with a high-frequency force.

high-frequency pulses that occurred twice a period. The spectrum of the pulses is widespread, and a component close to the pendulum frequency exists among far harmonics. This component swings the pendulum.

Considering friction and nonisochronism of the pendulum, we write its equation as follows:

$$u'' + 2\varepsilon\alpha u' + (1 - \varepsilon a^2)u = \varepsilon F g(t) \cos \nu t \tag{6.41}$$

Here ν is the frequency of the current, εF is the amplitude of the force, and $g(t)$ is a train of pulses shown in Figure 6.7 (black diamonds). We assume the following properties of the train:

- Its repetition rate is 2ω if ω is the pendulum frequency.
- Pulse duration depends on a pendulum velocity at the low position, $\tau \approx 2\delta/a\omega$, where $2\delta \ll a$ is the electromagnetic gap and a is the amplitude of a pendulum motion.
- The magnetic field and, therefore, the pulse shape is a smooth function. For concreteness, we set the shape $e^{-t^2/\tau^2} \cos \nu t$ for $\tau \ll \pi/\omega$, but the exact shape is unimportant though smoothness is significant.

The pulse train can be written as a Fourier series:

$$g(t) = \sum_{k=-\infty}^{\infty} c_k e^{i2k\omega t} \cos \nu t \qquad c_k = \frac{\delta}{\sqrt{\pi}\, a} e^{-k^2\delta^2/a^2} \tag{6.42}$$

Therefore, the equation for the AS takes the form

$$w'' + 2\varepsilon\alpha w' + (1 - \varepsilon a^2)w = \varepsilon F \sum_k c_k e^{i(2k\omega + \nu)t}$$

where the sum includes positive frequencies $2k\omega + \nu > 0$. The resonant component is defined by $2k\omega + \nu = \omega$, and therefore

$$\omega = \frac{\nu}{2n + 1} \quad \text{where } n = -k \quad \text{and} \quad \omega \approx 1 \tag{6.43}$$

Considering the resonant component only (factually, we use frequency separation again), we finally come to the equation in the form

$$w'' + 2\varepsilon\alpha w' + \Omega^2(a)w = \varepsilon F c_n e^{i\omega t} \quad \text{where } \Omega(a) = 1 - \varepsilon a^2/2 \tag{6.44}$$

which is the pendulum frequency dependent on its amplitude.

6.4.2 Resonant Oscillations—Modes

In general, exciting frequency ω differs from $\Omega(a)$, and nonlinear interaction is similar to that in Section 6.3 but more complicated as the force is amplitude-dependent. For each n, however, the *resonant amplitude* a_n exists that meets the condition $\Omega(a) = \omega$. According to (6.43) and (6.44), the resonant amplitude is given by

$$\sqrt{\varepsilon}a_n = \sqrt{2\left(1 - \frac{\nu}{2n+1}\right)} \tag{6.45}$$

For this amplitude, (6.44) takes the form

$$w'' + \omega^2 w = \varepsilon(-2\alpha w' + Fc_n e^{i\omega t}) \tag{6.46}$$

It has a harmonic solution $w = a_n e^{i(\omega t + \Phi)}$ if the right-hand side is zero. Setting it to zero and taking the c_n from (6.42), we come to the conditions for the amplitude and phase:

$$a^2 e^{n^2\delta^2/a^2} = \frac{F\delta}{2\sqrt{\pi}\alpha\omega} \qquad \Phi = -\frac{\pi}{2} \tag{6.47}$$

So the phase is shifted by $-\pi/2$, and differentiating the amplitude relation, we come to condition

$$\frac{\delta F}{F} = -2\left(\frac{n^2\delta^2}{a^2} - 1\right)\frac{\delta a}{a} \tag{6.48}$$

6.4.3 Amplitude Stabilization and Jumps

For $\nu \gg 1$, the number $n \approx \nu/2$ is so large that $n^2\delta^2/a^2 \gg 1$. Then according to (6.48), considerable variations of the force F cause small variations of the amplitude a. The amplitude is stabilized but slightly decreases when the force grows. Stabilization can be explained as follows.

The greater amplitude a, the shorter duration τ of exciting pulses, and the greater the exciting spectral component c_n for $n \gg 1$. Moreover, small variations of a and τ cause great variations of c_n because of the exponential dependence in (6.42). This comes from the smooth (analytical) magnetic field. So, great variations of the force F cause great variations of the exciting component c_n but small variations of the amplitude a.

Equation (6.47) is also applicable if the number n changes, and excitation happens with various spectral components c_n. So, the amplitude remains almost constant for a fixed n, but it jumps when n changes. Each n defines a resonant mode of oscillations.

Long-term interaction of two modes is impossible for this system. As we have seen, the interaction results in a varying amplitude that goes down to zero. At a small amplitude, exciting pulses disappear, and the pendulum remains at its lower position forever. Nevertheless, when changing the mode, short-term interaction befalls, and the "jump" is accomplished during half a cycle.

We have assumed that exciting pulses involve many periods of the feeding frequency ν. Otherwise, if the gap δ is so narrow that $n\delta/a \sim 1$, the factor a^2 dominates in (6.47), and $\delta a/a \sim \delta F/F$; then stabilization disappears.

6.4.4 On the Chronometer Precision

In watches and mechanical chronometers, the balance-wheel (oscillator) undergoes short blows from a crutch. In common watches, the blows occur once a period, but in chronometers, they occur once for about ten periods. For rare blows, one of the higher harmonics of a pulse train excites the balance-wheel. Now the pulses are low-frequency, without a carrier. If the pulses are smooth, however, higher harmonics depend on the oscillator amplitude in the same way as in (6.42). Therefore, the balance-wheel amplitude becomes independent of spring winding (like of the electromagnet current in Figure 6.7). In contrast with the pendulum considered, the balance-wheel is an *isochronic* oscillator. Nevertheless, the more stable amplitude, the more stable friction, and the more stable frequency of the balance-wheel (see Problem 6.3).

High chronometer precision is often explained in another way. It is considered that the longer period of blows $T_n = nT$, the smaller disturbances act on a free balance-wheel motion. However, for short, *rectangular* pulses of duration $\tau \ll T$, the n-th harmonic is given by

$$c_n = \frac{1}{\pi n} \sin\frac{\pi\tau}{T} = \frac{\tau}{nT} \quad \text{so that} \quad \frac{\delta c_n}{c_n} = \frac{\delta\tau}{\tau} = -\frac{\delta F}{F}$$

Therefore, variations of a force F are transformed into duration τ and into exciting amplitude c_n in the same extent as for $n = 1$ in common watches. *Smooth* pulses are needed for stabilization. This effect depends on *spectral properties* rather than a repetition rate of a pulse train.

6.5 Supplementary Problems

Problem 6.1

Using averaging, deduce adiabatic invariant (6.2) from (6.1).

Solution Multiplying (6.1) by u', we have

$$u''u' + \Omega^2 uu' = 0 \quad \text{or} \quad d(u')^2 + \Omega^2 du^2 = 0$$

Further, for a sinusoidal motion $u = a\cos(\Omega t + \Phi)$, we have $(u')^2 + \Omega^2 u^2 = \Omega^2 a^2$, and eliminating the $(u')^2$, we find

$$d(\Omega^2 a^2) - d(\Omega^2 u^2) + \Omega^2 du^2 = 0$$

Then using the identity $d(\Omega^2 u^2) = \Omega^2 du^2 + u^2 d\Omega^2$, we come to

$$d(\Omega^2 a^2) = \Omega^2 du^2 + u^2 d\Omega^2 - \Omega^2 du^2 = u^2 d\Omega^2$$

Finally, averaging $u^2(t) = a^2\cos^2(\Omega t + \Phi)$ over a period $2\pi/\Omega$, we obtain $\overline{u^2} = a^2/2$, and therefore

$$d(\Omega^2 a^2) = \frac{a^2}{2} d\Omega^2 \quad \text{or} \quad 2\Omega^2 da^2 + a^2 d\Omega^2 = 0$$

which is equivalent to $d(\Omega^2 a^4) = 0$ and $\Omega a^2 = \text{const}$.

Here, we assumed that modulation is slow and the quasi-sinusoidal motion can be averaged over a period while its amplitude and frequency remain constant. This approach is true for the first order only. As shown in Section 6.1, for nonoverlapping spectra, the momentum $K = \omega a^2$ is conserved in higher orders.

Problem 6.2

Solve (6.4) for a *complex* function $\Omega(\varepsilon t)$ and point out its relation to the WKB method [17].

Solution Taking ω as a new unknown, we denote

$$w(t) = \frac{\text{const}}{\sqrt{\omega}} e^{i\int \omega dt}$$

Then, we find

$$w' = \left(-\frac{\omega'}{2\omega} + i\omega\right) w \qquad w'' = \left(-\frac{\omega''}{2\omega} + \frac{3\omega'^2}{4\omega^2} - \omega^2\right) w$$

and substitution into (6.4) results in (6.8). Therefore, according to (6.9), $\omega = \Omega$ in the first order. So we obtain the WKB approximation

$$w(t) = \frac{\text{const}}{\sqrt{\Omega}} e^{i\int \Omega\, dt}$$

The amplitude now depends on the imaginary part of Ω, and the momentum K is not conserved.

Problem 6.3

Reduce the equation for the oscillator with friction

$$w'' + 2\alpha w' + \Omega^2 w = 0$$

to the form (6.4) taking $\alpha(t)$ and $\Omega(t)$ as arbitrary real functions. Also, if α is amplitude-dependent, show that, even for isochronic oscillators with $\Omega = const$, the oscillation frequency is amplitude-dependent. Therefore, chronometer precision is higher for stabilized amplitude as discussed in Section 6.4.4.

Solution For the new variable $\tilde{w}$ given by

$$w = \tilde{w} e^{-\int \alpha dt}$$

we have

$$a = \tilde{a} e^{-\int \alpha dt}$$

and the oscillator equation takes the form (6.4) with another frequency:

$$\tilde{w}'' + \tilde{\Omega}^2 \tilde{w} = 0 \quad \text{where } \tilde{\Omega}^2 = \Omega^2 - \alpha^2 - \alpha'$$

The subsequent solution is the same as in Section 6.1. This equation shows that the oscillation frequency is friction-dependent. Therefore, even for the isochronic balance-wheel with $\Omega = 1$, nonisochronism arises, and amplitude stabilization improves chronometer precision.

Problem 6.4

Using Problem 6.3, reduce generator equation (6.11) to the pendulum equation (6.4). Also, deduce (6.12) for momentum in another way.

Solution Rewriting (6.11) in the form

$$w'' - \varepsilon\gamma w' + (1 - \varepsilon\gamma')w = 0$$

and setting $\alpha = -\varepsilon\gamma/2$, we use Problem 6.3 to remove the term with w'. For this, we introduce the new oscillation $\bar{w}$ as follows:

$$w = \bar{w} \exp\left(\frac{\varepsilon}{2}\int \gamma\, dt\right)$$

and derive the equation of type (6.4):

$$\bar{w}'' + \left(1 - \frac{\varepsilon\gamma'}{2} - \frac{\varepsilon^2\gamma^2}{4}\right)\bar{w} = 0$$

Because γ is real, frequencies of w and $\bar{w}$ are identical. Also, since the pendulum momentum $\bar{K} = \omega\,\bar{a}^2$ is conserved, we have

$$K = \omega a^2 = \omega\,\bar{a}^2\, e^{\varepsilon\int\gamma\, dt} = \bar{K} e^{\varepsilon\int\gamma\, dt}$$

which is equivalent to the first equation (6.12).

Problem 6.5

Estimate spectral density for a shot noise and verify the value $N \sim 10^{-12}$ assumed in Section 6.2.3. Take typical values $I_0 = 10^{-2}\,A$, $\Gamma^2 = 10^{-3}$, and $f_0 = 5$ MHz for the current, depressing factor, and frequency, respectively.

Solution Random (white) shot current $i(t)$ is added to the mean current I_0 of an amplifier. Its spectral density is given by $N_i = \Gamma^2 e I_0$. Here $e = 1.6 \cdot 10^{-19}\,C$ is the electron charge and Γ^2 is the depressing factor resulting from a spatial charge.

Generator equation (6.20) has been written for dimensionless time $\omega_0 t$ and dimensionless current $I = 1$. Therefore, in agreement with Section 6.2.3, we come to the value:

$$N = \omega_0 \frac{N_i}{I_0^2} = 2\pi f_0 \frac{\Gamma^2 e}{I_0} \approx 5 \cdot 10^{-13}$$

Problem 6.6

For the pendulum between reflecting walls (Figure 6.5), estimate the time of cycle and show that time of linear swinging is greater than that of nonlinear interaction.

Solution Using (6.40), we find

$$\frac{da^2}{dt} = \frac{d(x^2+y^2)}{dt} = 2(xx' + yy')$$
$$= -(\Omega^2 - 1)xy - Fy + (\Omega^2 - 1)xy = -Fy$$

so that $$dt = -\frac{da^2}{Fy}$$

During linear swinging, $a^2 = y^2$ and $y < 0$; therefore, the time of swinging up to $a = 1$ is as follows:

$$T_1 = \frac{1}{F}\int_0^1 \frac{dy^2}{y} = \frac{2}{F}\int_0^1 dy = \frac{2}{F}$$

For nonlinear interaction, we replace the trajectory in Figure 6.6 by the ellipse

$$\frac{x^2}{a_{\min}^2} + y^2 = 1 \quad \text{so that} \quad a^2 = x^2 + y^2 = a_{\min}^2 + (1 - a_{\min}^2)y^2$$

where $a_{\min} \approx 1 - \Delta/\pi$ according to (6.38). Then, in the same way, we obtain

$$T_2 = \frac{1}{F}\int_0^1 \frac{(1 - a_{\min}^2)\,dy^2}{y} \approx \frac{4\Delta}{\pi F}$$

Finally, the total time of cycle is as follows:

$$T = 2(T_1 + T_2) \approx \frac{4}{F}\left(1 + \frac{2\Delta}{\pi}\right)$$

The time of the nonlinear interaction is less than that of linear swinging from $\Delta < \pi/2$.

Problem 6.7

In Section 6.2.2, the linear differential equation of the first order and the Bernoulli equation have been used. We now recall their solutions.

Linear Equation

$$u' + f(t)u + g(t) = 0$$

is solved with the substitution

$$u(t) = z(t)e^{-h(t)} \quad \text{where } h'(t) = f(t)$$

Then, $u' = (z' - fz)e^{-h}$, and the equation takes the form

$$[z' - f(t)z + f(t)z]e^{-h(t)} + g(t) = 0 \quad \text{or} \quad z' = -g(t)e^{h(t)}$$

Finally, we find

$$z(t) = -\int g(t)e^{h(t)}dt + C \qquad u(t) = \left\{-\int g(t)e^{h(t)}dt + C\right\}e^{-h(t)}$$

where C is the arbitrary constant.

Bernoulli Equation

$$u' + f(t)u + g(t)u^{\alpha} = 0$$

is reduced to the linear one by the substitution $u = z^{1/(1-\alpha)}$. Indeed, as

$$\frac{1}{1-\alpha} - 1 = \frac{\alpha}{1-\alpha}$$

we have

$$\frac{1}{1-\alpha}z^{\alpha/(1-\alpha)}z' + fz^{1/(1-\alpha)} + gz^{\alpha/(1-\alpha)} = 0$$

or

$$z' + (1-\alpha)f(t)z + (1-\alpha)g(t) = 0$$

References

[1] Andronov, A. A., S. E. Khaikin, and A. A. Vitt, *Theory of Oscillations*, Oxford, U.K.: Pergamon, 1966.

[2] Bogoliuboff, N. N., and Y. A. Mitropolskii, *Asymptotic Methods in the Theory of Nonlinear Oscillations,* New York: Gordon and Breach, 1961.

[3] Gilmore, R. J., and M. B. Steer, "Nonlinear Circuit Analysis Using the Method Harmonic Balance—A Review of the Art," *Int. J. Microwave Millimeter-Wave Computer-Aided Eng.*, Vol. 1, 1991, pp. 22–37 and 159–180.

[4] Hayashi, C., *Nonlinear Oscillations in Physical Systems,* New York: McGraw-Hill, 1964.

[5] Kevorkian, J., and J. D. Cole, *Perturbation Methods in Applied Mathematics,* New York: Springer-Verlag, 1981.

[6] Minorsky, N., *Nonlinear Oscillations,* Princeton, NJ: D. Van Nostrand, 1962.

[7] Nayfeh, A. H., and D. T. Mook, *Nonlinear Oscillations,* New York: Wiley, 1979.

[8] Hafner, E., "The Effects of Noise in Oscillators," *Proc. IEEE,* Vol. 54, 1966, pp. 179–198.

[9] Malahov, A. N., *Fluctuations in Auto-Oscillation Systems,* Moscow: Nauka-Press, 1968 (in Russian).

[10] Rytov, S. M., *Introduction to Statistical Radio-Physics,* Moscow: Nauka-Press, 1976 (in Russian).

[11] Van der Pol, B., "The Nonlinear Theory of Electrical Oscillations," *Proc. of the IRE,* Vol. 22, 1934, pp. 1051–1086.

[12] Buonomo, A., and C. D. Bello, "Asymptotic Formulas in Nearly Sinusoidal Nonlinear Oscillators," *IEEE Trans. Circuits and Systems,* pt. 1, Vol. 43, 1996, pp. 953–963.

[13] Wainstein, L. A., and D. E. Vakman, *Frequency Separation in Theory of Oscillations and Waves,* Moscow: Nauka-Press, 1983 (in Russian).

[14] Kulsrud, R. M., "Adiabatic Invariant of the Harmonic Oscillator," *Phys. Rev.*, Vol. 106, 1957, pp. 205–207.

[15] Vakman, D., "Dynamic Flicker Frequency Modulation and Noise Instability of Oscillators," *IEEE Trans. Circuits and Systems,* pt. 1, Vol. 41, 1994, pp. 321–325.

[16] Lewis, L. L., "An Introduction to Frequency Standards," *Proc. IEEE,* Vol. 79, 1991, pp. 927–935.

[17] Heading, J., *An Introduction to Phase-Integral Methods,* London: Methuen, New York: John Wiley, 1962.

7

Polyharmonic Oscillation Systems

In this chapter, we consider the *frequency separation procedure* (FSP) already used in Chapter 6. Studying the generator under a wideband additive noise (Section 6.2.3), we separated the resonant and outside noise components. Also, for the pendulum excited with a high-frequency force (Section 6.4.1), we separated the resonant component of a pulse train and neglected other components. Physically, this is understandable: the resonant component exerts a main influence on an oscillator.

The iterative FSP is more effective. It allows us to find out new spectral components appearing at each step and study their nonlinear interaction. This demonstrates important and interesting effects. So, nonisochronism of a pendulum and parametric resonances come from higher harmonics. For a generator, Van der Pol [1] considered the effect of harmonics as far back as the 1930s; he found out the static frequency correction discussed in Section 7.3.1. A general but very complicated method for harmonic interaction was developed in the 1940s [2] and used in many books [3,4]. The FSP simplifies this method and gives its physical interpretation [5]. We will show that the AS and nonoverlapping spectra are important for this procedure. The FSP also relates to perturbation and harmonic balance methods [6–8]. Some recent methods describe steady-state behavior of oscillators [9]. Defining the APF generally, however, the AS is applicable to transient regimes and modulation as well.

We generally study the oscillation equation in the form

$$u'' + u = \varepsilon f(u, u', t) \qquad \varepsilon \ll 1 \tag{7.1}$$

In Section 7.1, the iterative FSP will be developed for its solving. Then we apply the FSP to particular problems. Parametric resonances and wideband nonlinear generators are considered in this chapter. Electron motion and nonlinear waves will be considered in Chapters 8 and 9 with the same method.

7.1 Frequency Separation

As a preliminary example, we take the pendulum in Figure 7.1 for sizable angular deviations γ. Since the gravity force mg is vertical, the backward force is $-mg\sin\gamma$, and in a proper timescale, the pendulum obeys the nonlinear equation

$$\gamma'' + \sin\gamma = 0 \tag{7.2}$$

which can be written in the form (7.1) for $\gamma = \sqrt{\varepsilon}u$ and

$$f(u) = \frac{1}{\varepsilon}\left(u - \frac{\sin\sqrt{\varepsilon}u}{\sqrt{\varepsilon}}\right) = \frac{u^3}{3!} - \varepsilon\frac{u^5}{5!} + \cdots \tag{7.3}$$

First, we apply averaging method to this pendulum and show that it is efficient for the first order only.

Let us suppose that the motion is sinusoidal $u(t) = a\cos\omega t$ and its frequency ω is unknown. Using the Fourier series

$$\sin(\sqrt{\varepsilon}a\cos\omega t) = 2J_1(\sqrt{\varepsilon}a)\cos\omega t - 2J_3(\sqrt{\varepsilon}a)\cos 3\omega t + \cdots$$

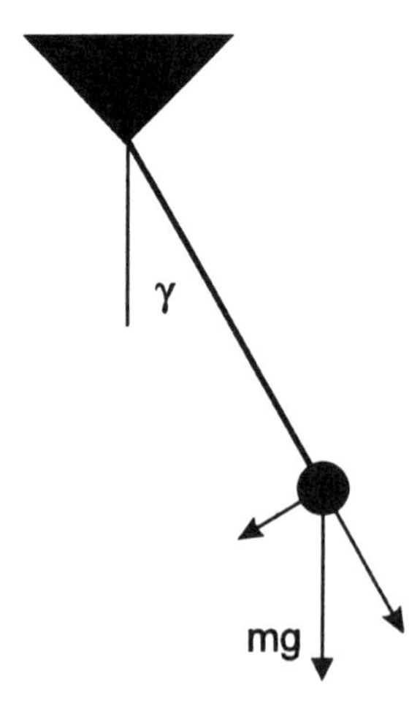

Figure 7.1 Nonlinear pendulum.

where J_n are Bessel functions, we obtain from (7.2):

$$u'' + \frac{2J_1(\sqrt{\varepsilon}a)}{\sqrt{\varepsilon}a}u = \frac{2}{\sqrt{\varepsilon}}\{J_3(\sqrt{\varepsilon}a)\cos 3\omega t - \cdots\}$$

We reach a conflict. For $u = a\cos\omega t$, the left-hand side contains the frequency ω, whereas the right-hand side contains the frequencies $3\omega,\ 5\omega, \ldots$. Therefore, equality cannot be held. Nevertheless, neglecting nonresonant higher harmonics or averaging the equation after multiplication by $\cos\omega t$, we find the following equation:

$$u'' + \omega^2 u = 0 \quad \text{where } \omega^2 = \frac{2J_1(\sqrt{\varepsilon}a)}{\sqrt{\varepsilon}a} = 1 - \frac{\varepsilon a^2}{8} + \frac{\varepsilon^2 a^4}{192} - \cdots \tag{7.4}$$

Unfortunately, this simple approach is right for the first order only, but the second-order term with ε^2 should be amended. This is shown with exact solution in Problem 7.1 and with the FSP below. The cause of inaccuracy is clear: higher harmonics distort the frequency. With averaging, we simplify a problem by removing higher harmonics, but we lose accuracy.

7.1.1 Iterations and Frequency Separation

Now, considering (7.1) with the function $f(u)$ given in (7.3), we seek its solution in the form

$$u = \sum_k u_k(t) \qquad u_k = \mathrm{Re}\, w_k \qquad w_k = z_k e^{ikt} \qquad k = 0, 1, 2\ldots \tag{7.5}$$

The frequency ω differs from $\omega_0 = 1$, and for a *periodic* pendulum motion, we assume that

$$z_k(t) = Z_k e^{ik(\omega-1)t} \qquad w_k(t) = Z_k e^{ik\omega t} \qquad Z_k = \text{const.}$$

Then (7.5) is the Fourier series for $u(t)$.

Let $u^{(n)}(t)$ be the n-order approximation, which is the solution of (7.1) in the form (7.5) accomplished up to ε^n. This means that the harmonics smaller than ε^n are eliminated, and the amplitudes z_k are computed up to ε^n. Then we write the next approximation in the form

$$u^{(n+1)}(t) = \tilde{u}^{(n)}(t) + \varepsilon^{n+1}\beta(t) \tag{7.6}$$

Here $\bar{u}^{(n)}$ contains the same harmonics as $u^{(n)}$, but their amplitudes are defined more accurately, up to ε^{n+1}. The $\beta(t)$ contains new harmonics of amplitudes ε^{n+1} neglected in the n-order. Note that spectra of $\bar{u}^{(n)}$ and β are nonoverlapping.

Substituting (7.6) into (7.1), we have

$$\frac{d^2 u^{(n+1)}}{dt^2} + u^{(n+1)} = \varepsilon f(\bar{u}^{(n)} + \varepsilon^{n+1}\beta) = \varepsilon f(\bar{u}^{(n)}) + \varepsilon^{n+2} f'(\bar{u}^{(n)})\beta + \cdots$$

and neglecting the addends with ε^{n+2} and up, we come to the iterative equation

$$\frac{d^2 u^{(n+1)}}{dt^2} + u^{(n+1)} = \varepsilon f(\bar{u}^{(n)}) \tag{7.7}$$

which defines the next approximation from the preceding one. We emphasize that (7.7) must be solved again at each step. We should not take the $u^{(n)}$ from the preceding step because the amplitudes z_k are to be updated. Since z_k are *complex* amplitudes, frequencies are also updated at each step. Preceding solution defines the *spectral composition* of $\bar{u}^{(n)}$ in the right-hand part of (7.7), but the amplitudes are to be found from this equation. Nevertheless, simplifying notations, we write $u^{(n)}$ instead of $\bar{u}^{(n)}$, implying a new solution at each step.

Further, for the known spectral composition, we *separate frequencies* in (7.7) and extract individual equations for each harmonic $u_k(t)$ composing the solution (7.5). For a *periodic* pendulum motion, the separation is trivial: using a Fourier series, we equate harmonic components of equal frequencies from both sides of (7.7).

However—and this point is fundamental—frequency separation is applicable generally for quasiharmonic components of nonoverlapping spectra. Although their spectra are of finite bands, equating the components from both sides of (7.7) is right if they belong to the same separable spectral band. Moreover, since both sides of (7.7) must contain the same spectral components, from the known composition of $\bar{u}^{(n)}$ and $f(\bar{u}^{(n)})$, we find the components existing in $u^{(n+1)}$. Thus, frequency separation assumes nonoverlapping spectra, and the equated components relate to analytic signals. Sometimes, spectra of certain components are overlapping and inseparable. Then we come to new interesting phenomena such as parametric resonances (Section 7.2). We will detail the FSP below.

7.1.2 Nonlinear Pendulum

First-Order Solution

Now we apply frequency separation to the pendulum in Figure 7.1. For $\varepsilon = 0$, equation (7.1) $u'' + u = 0$ has a harmonic solution with $\omega = 1$. Therefore, using notations (7.5), we write $u^{(0)} = u_1$, and the first-order iterative equation (7.7) takes the form

$$\frac{d^2 u^{(1)}}{dt^2} + u^{(1)} = \frac{\varepsilon}{3!} u_1^3 \tag{7.8}$$

Here, the first addend in (7.3) is only considered since the second addend is of the order ε^2.

Further, we transfer to the AS by setting

$$u_1 = \frac{w_1 + w_1^*}{2} \qquad u_1^3 = \frac{w_1^3 + 3w_1^2 w_1^* + 3w_1 w_1^{*2} + w_1^{*3}}{8}$$

and eliminating the addends at negative frequencies (Property 3, Section 2.2). The components $w_1^* = z_1^* e^{-it}$, $w_1^{*3} = z_1^{*3} e^{-i3t}$, and $w_1 w_1^{*2} = |z_1|^2 z_1^* e^{-it}$ are at negative frequencies $\omega = -1$ and $\omega = -3$ and should be eliminated. Then we obtain from (7.8)

$$\frac{d^2 w^{(1)}}{dt^2} + w^{(1)} = \frac{\varepsilon}{24}(3a^2 w_1 + w_1^3) \qquad a^2 = w_1 w_1^* \tag{7.9}$$

The two addends in the right-hand side of (7.9) are $w_1 = z_1 e^{it}$ and $w_1^3 = z_1^3 e^{3it}$. They belong to separate frequency bands around $\omega = 1$ and $\omega = 3$, and the same spectral components must be in the left-hand side. Therefore, using notations (7.5), we set $w^{(1)} = w_1 + w_3$, and w_3 as a new spectral component appearing in the first order. Separating frequencies, we come to individual equations for w_1 and w_3:

$$\begin{aligned} w_1'' + \left(1 - \frac{\varepsilon}{8} a^2\right) w_1 &= 0 \\ w_3'' + w_3 &= \frac{\varepsilon}{24} w_1^3 \end{aligned} \tag{7.10}$$

The equation for w_1 defines the updated frequency

$$\omega^2 = 1 - \frac{\varepsilon}{8} a^2 \tag{7.11}$$

and the first-order correction $-\varepsilon a^2/8$ is the same as in (7.4). This frequency dependence has been assumed in Chapter 6 for nonisochronic oscillators. Clearly, it comes from the term $a^2 w_1 = w_1^2 w_1^*$ in (7.9), which is the nonlinear self-action of a main oscillation.

The second equation (7.10) defines the third harmonic w_3. In general, its solution contains a *proper* motion at $\omega = 1$ and a *forced* motion at $\omega = 3$. However, the first harmonic is included in w_1, and we keep a forced motion only. Then, setting $w_3 = z_3 e^{3it}$ and differentiating, we find

$$-9w_3 + w_3 = \frac{\varepsilon}{24} w_1^3 \quad \text{so that } w_3 = -\frac{\varepsilon}{192} w_1^3 \tag{7.12}$$

Second-Order Solution

Relations (7.11) and (7.12) define the first-order solution. The second-order solution can be found in the same way, but now we know that $u^{(1)} = u_1 + u_3$ where $u_3 \sim \varepsilon$. Keeping the terms up to ε^2 in the iterative equation (7.7), we obtain

$$\frac{d^2 u^{(2)}}{dt^2} + u^{(2)} = \frac{\varepsilon}{3!}(u^{(1)})^3 - \frac{\varepsilon^2}{5!}(u^{(1)})^5 = \frac{\varepsilon}{6}(u_1^3 + 3u_1^2 u_3) - \frac{\varepsilon^2}{120} u_1^5$$

Then, setting $u_1 = (w_1 + w_1^*)/2$, we expand u_1^3 and u_1^5 and eliminate the components at negative frequencies. We then come to an equation for the AS as follows:

$$\frac{d^2 w^{(2)}}{dt^2} + w^{(2)} = \frac{\varepsilon}{6}\left[\frac{w_1^3 + 3w_1^2 w_1^*}{4} + 3u_1^2 w_3\right] - \frac{\varepsilon^2}{1920}(w_1^5 + 5w_1^4 w_1^* + 10 w_1^3 w_1^{*2})$$

We have also taken into account that the spectrum of u_1^2 is lower than that of u_3 and obtained the term $u_1^2 w_3$ using Bedrosian's theorem.

Let us consider the frequencies in the right-hand part. The $w_1^2 w_1^* = a^2 w_1 = a^2 z_1 e^{it}$ is the first harmonic at $\omega = 1$. The $w_1^3 = z_1^3 e^{i3t}$ is the third harmonic at $\omega = 3$. The $u_1^2 w_3 = (w_1^2 w_3 + 2a^2 w_3 + w_1^{*2} w_3)/4$ contains the first harmonic $w_1^{*2} w_3 = z_1^{*2} z_3 e^{it}$, the third harmonic $a^2 w_3 = a^2 z_3 e^{i3t}$, and the fifth harmonic $w_1^2 w_3 = z_1^2 z_3 e^{i5t}$. Also, all three harmonics are in the sum $w_1^5 + 5w_1^4 w_1^* + 10 w_1^3 w_1^{*2} = z_1^5 e^{i5t} + 5a^2 z_1^3 e^{i3t} + 10 a^4 z_1 e^{it}$. Therefore, we set $w^{(2)} = w_1 + w_3 + w_5$ in the left-hand side, and the w_5 is a new spectral component appearing in the second order. For frequency separation, we collect the components of each frequency and obtain individual equations as follows:

$$\begin{aligned}
w_1'' + w_1 &= \frac{\varepsilon}{8}(a^2 w_1 + w_1^{*2} w_3) - \frac{\varepsilon^2}{192} a^4 w_1 \\
w_3'' + w_3 &= \varepsilon\left(\frac{1}{24} w_1^3 + a^2 w_3\right) - \frac{\varepsilon^2}{384} a^2 w_1^3 \\
w_5'' + w_5 &= \frac{\varepsilon}{8} w_1^2 w_3 - \frac{\varepsilon^2}{1920} w_1^5
\end{aligned} \tag{7.13}$$

This system defines w_1, w_3, and w_5 up to the second order ε^2, and it should be solved again, with no preceding results being used. However, when substituting into the first equation, the w_3 should be found up to the first order ε since the additional factor ε exists in the right-hand side. For this accuracy, we neglect the terms $\varepsilon a^2 w_3$ and $\varepsilon^2 a^2 w_1^3$ (both of the order ε^2) in the second equation (7.13). Then, the equation for w_3 is the same as in (7.10), and its solution is given in (7.12). Substituting this solution into the first equation (7.13), we obtain

$$w_1'' + \omega^2 w_1 = 0 \quad \text{where } \omega^2 = 1 - \frac{\varepsilon}{8}a^2 + \frac{3\varepsilon^2}{512}a^4 \tag{7.14}$$

This frequency differs from (7.4) in the second order but agrees with exact solution in Problem 7.1. The distinction comes from the term $w_1^{*2} w_3$ in the first equation (7.13). This term describes a contribution into w_1 from nonlinear interaction of the first and third harmonics. If w_3 is removed with averaging, the frequency is the same as in (7.4).

Knowing the w_1, we find w_3 and w_5 from (7.13) up to ε^2 (again keeping forced oscillations only):

$$w_3 = -\frac{\varepsilon}{192}\left(1 - \frac{3\varepsilon a^2}{16}\right) w_1^3 \qquad w_5 = \frac{\varepsilon^2}{20480} w_1^5$$

Clearly, continuing iterations, we can find higher corrections to frequency and higher harmonics. Although these small corrections may cost us too much effort, we now know how to solve oscillation problems better than with averaging.

7.2 Parametric Resonances

We return to the pendulum of varying length (Figure 6.1), which obeyed the equation

$$u'' + \Omega^2(t) u = 0$$

but now we assume that the carriage moves *fast*, and its frequency is higher than the pendulum frequency. For simplicity, we set

$$\Omega^2(t) = 1 - \varepsilon \cos pt \quad \text{where } \varepsilon \ll 1 \text{ and } p > 1$$

and come to Mathieu equation

$$u'' + u = \varepsilon u \cos pt \tag{7.15}$$

For fast modulation, the pendulum displays very different behavior. Instead of the conserved momentum $K = \omega a^2$, parametric resonances appear, and the pendulum is swinging up to unlimited amplitude. Although the two regimes obey the same real equation (7.15), complex equations for the AS are different, which results in contrasting physical phenomena.

7.2.1 Frequency Separation and the AS

For $\varepsilon = 0$, (7.15) has a harmonic solution of unit frequency. So, we set $u^{(0)} = u_1$, and the first-order iterative equation (7.7) takes the form

$$\frac{d^2 u^{(1)}}{dt^2} + u^{(1)} = \varepsilon u_1 \cos pt$$

Let us apply the HT to this equation. Since $p > 1$ is higher than the frequency $\omega = 1$ of u_1, using Bedrosian's theorem, we take u_1 out of the HT and obtain an equation for the AS as follows:

$$\frac{d^2 w^{(1)}}{dt^2} + w^{(1)} = \varepsilon u_1 e^{ipt} = \frac{\varepsilon}{2}(w_1 e^{ipt} + w_1^* e^{ipt}) \tag{7.16}$$

where we set $u_1 = (w_1 + w_1^*)/2$. This AS equation differs from (6.4) for the same pendulum. In Section 6.1.1, using Bedrosian's theorem for slow modulation, we took out the cos pt, but for fast modulation, we take out the u_1 and come to another complex equation.

The right-hand side of (7.16) contains the frequencies $p + 1$ and $p - 1$, and the same frequencies are found in the left-hand side. So, we set $w^{(1)} = w_1 + w_{p+1} + w_{p-1}$, and after frequency separation, obtain three equations:

$$\begin{aligned} w_1'' + w_1 &= 0 \\ w_{p-1}'' + w_{p-1} &= \frac{\varepsilon}{2} w_1^* e^{ipt} \\ w_{p+1}'' + w_{p+1} &= \frac{\varepsilon}{2} w_1 e^{ipt} \end{aligned} \tag{7.17}$$

The first equation defines oscillations of unit frequency $\omega = 1$, whereas the two others define oscillations at $p - 1$ and $p + 1$. So, fast modulation splits the spectrum into three components.

First Resonance

There is an important exception, however. For $p \approx 2$ and $p - 1 \approx 1$, the first and second equations (7.17) are both for $\omega \approx 1$ and cannot be separated. Then we come to two equations:

$$\begin{aligned} w_1'' + w_1 &= \frac{\varepsilon}{2} w_1^* e^{ipt} \\ w_3'' + w_3 &= \frac{\varepsilon}{2} w_1 e^{ipt} \end{aligned} \tag{7.18}$$

and the first one results in the resonant excitation.

For solving this equation, we introduce the complex amplitude $z = x + iy$ by setting $w_1 = z(t)e^{ipt/2}$. Then the equation takes the form

$$z'' + ipz' + \left(1 - \frac{p^2}{4}\right) z = \frac{\varepsilon}{2} z^*$$

and separating its real and imaginary parts, we come to the homogeneous system for quadratures:

$$\begin{cases} x'' - py' + \left(1 - \dfrac{p^2}{4} - \dfrac{\varepsilon}{2}\right) x = 0 \\ y'' + px' + \left(1 - \dfrac{p^2}{4} + \dfrac{\varepsilon}{2}\right) y = 0 \end{cases}$$

Its solution is $x = C_1 e^{\lambda t}$, $y = C_2 e^{\lambda t}$, and for $\lambda > 0$, oscillations are growing. Substituting these x and y, we come to the algebraic system for C_1 and C_2:

$$\begin{cases} \left(\lambda^2 + 1 - \dfrac{p^2}{4} - \dfrac{\varepsilon}{2}\right) C_1 - p\lambda C_2 = 0 \\ p\lambda C_1 + \left(\lambda^2 + 1 - \dfrac{p^2}{4} + \dfrac{\varepsilon}{2}\right) C_2 = 0 \end{cases}$$

which has a solution if its determinant is zero. So we come to the characteristic equation for the increment λ:

$$\begin{vmatrix} \lambda^2 + 1 - \dfrac{p^2}{4} - \dfrac{\varepsilon}{2} & -p\lambda \\ p\lambda & \lambda^2 + 1 - \dfrac{p^2}{4} + \dfrac{\varepsilon}{2} \end{vmatrix} = \lambda^4 + 2\left(1 + \frac{p^2}{4}\right)\lambda^2 + \left(1 - \frac{p^2}{4}\right)^2 - \frac{\varepsilon^2}{4} = 0$$

For λ^2 to be positive, the free term $(1 - p^2/4)^2 - \varepsilon^2/4$ must be negative, and we come to the following resonant condition:

$$\left|1 - \frac{p^2}{4}\right| < \frac{\varepsilon}{2} \quad \text{or} \quad \delta p = |p - 2| < \frac{\varepsilon}{2} \tag{7.19}$$

which defines the *zone of excitation* where oscillations are growing. For the middle of the zone $p = 2$, the increment $\lambda = \varepsilon/4$ is maximal, and the oscillation grows as follows:

$$w_1 = Ce^{\varepsilon t/4+it}$$

We can see from the first equation (7.18) that the resonance comes from interaction between w_1^* and modulating frequency $p \approx 2$.

The second equation (7.18) defines the third harmonic:

$$w_3 = -\frac{\varepsilon}{16}w_1e^{2it} = -\frac{\varepsilon}{16}Ce^{\varepsilon t/4+3it}$$

which is smaller than w_1 but also growing.

Second Resonance

For $p \neq 2$, equations (7.17) are valid, and $u^{(1)} = u_1 + u_{p+1} + u_{p-1}$. Therefore, the second-order iterative equation (7.7) takes the form

$$\frac{d^2u^{(2)}}{dt^2} + u^{(2)} = \varepsilon(u_1 + u_{p+1} + u_{p-1})\cos pt \tag{7.20}$$

Frequencies of u_1 and u_{p-1} are lower than p, but of u_{p+1} is higher than p. Therefore, using Bedrosian's theorem, we come to equation for the AS as follows:

$$\frac{d^2w^{(2)}}{dt^2} + w^{(2)} = \varepsilon(u_1 + u_{p-1})e^{ipt} + \varepsilon w_{p+1}\cos pt \tag{7.21}$$

Setting $u_1 = (w_1 + w_1^*)/2$ and similarly for u_{p-1} and $\cos pt$, we then find that the frequencies 1, $p-1$, $p+1$, $2p-1$, and $2p+1$ exist in the right-hand side. The same frequencies are found in the left-hand side, and separating them, we come to the second-order system as follows:

$$
\begin{aligned}
w_1'' + w_1 &= \frac{\varepsilon}{2}\left(w_{p-1}^* e^{ipt} + w_{p+1} e^{-ipt}\right) \\
w_{p-1}'' + w_{p-1} &= \frac{\varepsilon}{2} w_1^* e^{ipt} \\
w_{p+1}'' + w_{p+1} &= \frac{\varepsilon}{2} w_1 e^{ipt} \\
w_{2p-1}'' + w_{2p-1} &= \frac{\varepsilon}{2} w_{p-1} e^{ipt} \\
w_{2p+1}'' + w_{2p+1} &= \frac{\varepsilon}{2} w_{p+1} e^{ipt}
\end{aligned}
\tag{7.22}
$$

The second resonance appears at $p \approx 1$ when $2p - 1 \approx 1$, and equations for w_1 and w_{2p-1} are inseparable. Then, the system (7.22) takes the form

$$
\begin{aligned}
w_1'' + w_1 &= \frac{\varepsilon}{2}\left(w_0^* e^{ipt} + w_2 e^{-ipt} + w_0 e^{ipt}\right) \\
w_0'' + w_0 &= \frac{\varepsilon}{2} w_1^* e^{ipt} \\
w_2'' + w_2 &= \frac{\varepsilon}{2} w_1 e^{ipt} \\
w_3'' + w_3 &= \frac{\varepsilon}{2} w_2 e^{ipt}
\end{aligned}
\tag{7.23}
$$

From the second and third equations, we obtain (again up to the first order with respect to ε)

$$
w_0 = \frac{\varepsilon}{2} w_1^* e^{ipt} \qquad w_2 = -\frac{\varepsilon}{6} w_1 e^{ipt}
$$

Then, substituting into the first equation, we come to the resonant equation

$$
w_1'' + \left(1 - \frac{\varepsilon^2}{6}\right) w_1 = \frac{\varepsilon^2}{4} w_1^* e^{2ipt} \tag{7.24}
$$

which can be solved in the same way as (7.18). Finally, the zone of excitation and maximal increment are given by

$$
-\frac{5\varepsilon^2}{24} < \delta p < \frac{\varepsilon^2}{24} \qquad \lambda = \frac{\varepsilon^2}{8}
$$

Compared with the first resonance, the zone is narrower and the increment is smaller—both of the order ε^2. Note also that the complex conjugate AS w_1^* interacts in (7.24) with the second harmonic of modulation. Therefore, like for the first order, resonance appears from parametric excitation at $\omega = 2$.

Other equations (7.23) show that the resonance is accompanied with the second and third harmonics plus the slow drift w_0. As in the synchronous detector, the drift comes from multiplication of oscillations with equal frequencies.

Third Resonance

A resonance appears each time when a combinative frequency is $\omega = 1$. We have seen this for the frequencies $p-1$ and $2p-1$ at $p = 2$ and $p = 1$, respectively. Also, the frequencies $3p-1$, $4p-1, \ldots,$ when appearing in higher orders, produce resonances. For the third order, the resonance occurs at $p = 2/3$, when $3p-1 = 1$ and also $2p-1 = |p-1| = 1/3$, $p+1 = 5/3$, and $2p+1 = 7/3$. Therefore, according to (7.22), the iterative equation (7.7) takes the form

$$\frac{d^2u^{(3)}}{dt^2} + u^{(3)} = \varepsilon(u_1 + u_{1/3} + u_{5/3} + u_{7/3}) \cos pt$$

Here the frequencies of u_1, $u_{5/3}$, and $u_{7/3}$ are higher, but of $u_{1/3}$ is lower than $p = 2/3$. Therefore, after coming to the AS and frequency separation, we obtain the third-order resonant system as follows:

$$\begin{aligned}
w_1'' + w_1 &= \frac{\varepsilon}{2}(w_{1/3}e^{ipt} + w_{5/3}e^{-ipt}) \\
w_{1/3}'' + w_{1/3} &= \frac{\varepsilon}{2}(w_1 e^{-ipt} + w_{1/3}^* e^{ipt}) \\
w_{5/3}'' + w_{5/3} &= \frac{\varepsilon}{2}(w_1 e^{ipt} + w_{7/3}e^{-ipt}) \\
w_{7/3}'' + w_{7/3} &= \frac{\varepsilon}{2} w_{5/3}e^{ipt} \\
w_3'' + w_3 &= \frac{\varepsilon}{2} w_{7/3}e^{ipt}
\end{aligned} \tag{7.25}$$

Up to the second order, the $w_{1/3}$ and $w_{5/3}$ can be found from the second and third equations in the forms

$$w_{1/3} = \frac{9\varepsilon}{16}\left(w_1 e^{-ipt} + \frac{9\varepsilon}{16} w_1^* e^{2ipt}\right) \qquad w_{5/3} = -\frac{9\varepsilon}{32} w_1 e^{ipt}$$

and the first equation becomes analogous to (7.18) and (7.24):

$$w_1'' + \left(1 - \frac{9\varepsilon^2}{64}\right) w_1 = \frac{81\varepsilon^3}{512} w_1^* e^{3ipt} \qquad (7.26)$$

Its solution found in the same way determines the zone of excitation and the increment of the order ε^3:

$$\left|\delta p + \frac{3\varepsilon^2}{64}\right| < \frac{27\varepsilon^3}{512} \qquad \lambda = \frac{81\varepsilon^3}{1024}$$

So, the higher the order of a parametric resonance, the narrower the zone of excitation and the slower the growing. For each order, interaction between w_1^* and the modulating harmonic at $np \approx 2$ produces the resonance at $\omega = 1$.

7.2.2 Bifurcations and the AS

Parametric oscillations display *bifurcations*—sharp changes of system behavior while its parameter is slowly varying. Depending on the modulating frequency p, three modes are feasible:

- *Adiabatic mode*—slow variations of frequency and amplitude with the conserved momentum ωa^2;
- *Modulation mode*—basic harmonic oscillation at $\omega = 1$, with side components of combinative frequencies;
- *Resonant mode*—growing oscillations with growing harmonics.

All the modes are observable for the same pendulum and obey the same real equation of motion. We cannot comprehend from this equation, however, when and why the modes change each other. On the other hand, complex equations for the AS are different for each mode, and Bedrosian's theorem explains why pendulum behavior is different.

7.3 Wideband Generator Theory

In Section 6.2, we considered the generator with a narrowband amplifier assuming that its current $I(t) = k(a)u(t)$ follows the input voltage $u(t)$ and the gain-factor $k(a)$ depends on the input amplitude. Dynamic frequency variations and flicker instability have been studied for this generator.

Other nonlinearities appear in *wideband* amplifiers. Van der Pol has considered a *cubic nonlinearity* that can be written as

$$I(u) = 2u - \frac{4}{3}u^3 \tag{7.27}$$

For operational amplifiers and flip-flops, *step nonlinearity* is typical:

$$I(u) = \text{sgn}(u) = \begin{cases} -1 & \text{for } u < 0 \\ 1 & \text{for } u > 0 \end{cases} \tag{7.28}$$

For both, a sinusoidal input voltage is distorted in the amplifier, and higher harmonics appear. Their interaction may result in additional effects missed in Section 6.2, and the theory has to be revised.

If, as before, $\varepsilon = 1/Q \ll 1$ is a small parameter and $m(\mu t)$ is slow modulation, the generator equation

$$u'' + u = \varepsilon \frac{d}{dt}[I(u)m(\mu t) - u] \tag{7.29}$$

is the special case of (7.1), and our iterations are applicable. Using the FSP up to the second order, we will come to the following:

- *Dynamic* frequency fluctuations remain the same as for narrowband generators, and according to Section 6.2, flicker instability dominates over other noise effects.
- For cubic nonlinearity, interaction of higher harmonics results in the additional *static* frequency correction. The correction is negligible for high-quality resonators, however.
- For step nonlinearity, higher harmonics do not interact with each other, and dynamic frequency distortion is the only source of instability.

Static frequency correction was known as early as the 1930s [1]. It has also been reasserted with many modern methods [8,9], but dynamic distortions have not been discovered. Since static correction is small, frequency instability of generators has not been clarified.

7.3.1 Cubic Nonlinearity

First-Order Solution For $\varepsilon = 0$, (7.29) takes the form $u'' + u = 0$, so that $u^{(0)} = u_1$. Therefore, for the cubic characteristic (7.27), iterative equation (7.7) is as follows:

$$\frac{d^2u^{(1)}}{dt^2} + u^{(1)} = \varepsilon\frac{d}{dt}\left[(2m-1)u_1 - \frac{4m}{3}u_1^3\right] \quad (7.30)$$

For slow modulation, we take $m(\mu t)$ out of the HT, and then, in the same way as (7.9), the equation for the AS takes the form

$$\frac{d^2w^{(1)}}{dt^2} + w^{(1)} = \varepsilon\frac{d}{dt}\left[(2m-1-ma^2)w_1 - \frac{m}{3}w_1^3\right] \quad (7.31)$$

Separating the first and third harmonics, we come to the first-order system:

$$\begin{aligned} w_1'' + w_1 &= \varepsilon\frac{d}{dt}[(2m-1-ma^2)w_1] \\ w_3'' + w_3 &= -\varepsilon\frac{d}{dt}\left(\frac{m}{3}w_1^3\right) \end{aligned} \quad (7.32)$$

The first equation (7.32) is the same as (6.11) but with another slow function $\gamma = 2m - 1 - ma^2$. In Section 6.2, this equation has led to dynamic frequency distortions (6.13) independent of a particular function γ. So, the same dynamic frequency fluctuations can be expected for the generator under consideration. This conclusion may be premature, however, because frequency fluctuations are of the second order, but we have obtained only the first-order equations as yet.

Keeping the forced oscillation only and neglecting derivatives of slow functions, we find the third harmonic from the second equation (7.32):

$$w_3 = \frac{\varepsilon}{24}\frac{d}{dt}(mw_1^3) \approx \frac{i\varepsilon}{8}mw_1^3 \quad (7.33)$$

The factor i comes from differentiation $dw_1^3/dt = 3w_1^2w_1' \approx 3iw_1^3$, where $w_1 = z_1e^{it}$. This factor shows the phase shift by $\pi/2$ between the first and third harmonics. It also leads to the static frequency correction (see (7.38) below).

Second-Order Solution Now, for $u^{(1)} = u_1 + u_3$ where $u_3 \sim \varepsilon$, iterative equation (7.7) takes the form (we keep the terms up to ε^2):

$$\frac{d^2u^{(2)}}{dt^2} + u^{(2)} = \varepsilon\frac{d}{dt}\left[(2m-1)(u_1+u_3) - \frac{4m}{3}(u_1^3 + 3u_1^2u_3)\right] \quad (7.34)$$

Then, after transfer to the AS and frequency separation, we come to individual equations for w_1, w_3, and w_5:

$$\begin{aligned} w_1'' + w_1 &= \varepsilon \frac{d}{dt} [(2m - 1 - ma^2) w_1 - m w_1^{*2} w_3] \\ w_3'' + w_3 &= \varepsilon \frac{d}{dt} \left[(2m - 1 - 2ma^2) w_3 - \frac{m}{3} w_1^3 \right] \\ w_5'' + w_5 &= \varepsilon \frac{d}{dt} (m w_1^2 w_3) \end{aligned} \tag{7.35}$$

As usual, for substitution into the first equation, we find the w_3 up to the first order ε. Therefore, in the right-hand part of the second equation, we neglect the second-order addend with w_3. Then the equation is the same as in (7.32), and its solution is given by (7.33). Substituting w_3 from (7.33), we obtain

$$w_1'' + w_1 = \varepsilon \frac{d}{dt} \left[\left(2m - 1 - ma^2 - i \frac{\varepsilon}{8} m^2 a^4 \right) w_1 \right] \tag{7.36}$$

which contains a small *imaginary* correction to $\gamma = 2m - 1 - ma^2$.

This correction results in a new effect. Moving the term with ε^2 into the left-hand side, we neglect derivatives of slow factors (with the total error of ε^3) and come to the equation in the form

$$w_1'' + i \frac{\varepsilon^2}{8} m^2 a^4 w_1' + w_1 = \varepsilon \frac{d}{dt} [(2m - 1 - ma^2) w_1] \tag{7.37}$$

Then, using Problem 6.3, we eliminate the addend with w_1'. For this, we introduce the new unknown $\tilde{w}_1$, which differs from w_1 by the frequency correction $\Delta\omega$:

$$w_1 = \tilde{w}_1 \exp\left(-i \frac{\varepsilon^2}{16} \int m^2 a^4 dt \right) \qquad \Delta\omega = -\frac{\varepsilon^2}{16} m^2 a^4 \tag{7.38}$$

After substitution into (7.37), the equation for $\tilde{w}_1$ takes the form (again with an error of ε^3):

$$\tilde{w}_1'' + \tilde{w}_1 = \varepsilon \frac{d}{dt} [(2m - 1 - ma^2) \tilde{w}_1]$$

This is the same equation as (7.32) or (6.11) but for the *second* order with respect to ε. Therefore, the $\tilde{w}_1$ is a signal with dynamic frequency fluctuations (6.13) discussed in Section 6.2. We recall that dynamic fluctuations dependent on the second derivative $a''(t)$ come from amplitude modulation to support the phase balance in the generator.

According to (7.38), however, the signal w_1 of the generator also contains *static* frequency correction $\Delta\omega$ dependent on the amplitude $a(t)$ but not its derivative. Similar to the pendulum in Section 7.1.2, this correction comes from nonlinear interaction of harmonics—namely, from the term $w_1^{*2}w_3$ in the first equation (7.35) that shows interaction of the first and third harmonics.

So, two second-order effects produce frequency fluctuations: nonlinear interaction of harmonics and frequency modulation resulted from amplitude variations to support the phase balance. Steady-state methods [2,6–9] describe the static correction only, but the AS and FSP clarify both effects. Being dependent on ε^2, static correction $\Delta\omega$ is small for high-quality resonators, and dynamic fluctuations dominate.

7.3.2 Step Nonlinearity

According to (7.28), the amplifier current $I(t)$ is a rectangular wave as illustrated in Figure 7.2. It can be written as a Fourier series:

$$I(t) = \cos\omega t - \frac{1}{3}\cos 3\omega t + \frac{1}{5}\cos 5\omega t - \cdots \tag{7.39}$$

In the figure, the input voltage is sinusoidal. The rectangular current, however, remains the same for signals with small odd harmonics (zero-crossings are at the same time points). Coming to the AS in (7.39), we replace $\cos k\omega t$ by $e^{ik\omega t}$, and the AS of the rectangular current takes the form

$$w_I(t) = \sum_k (-1)^{\frac{k-1}{2}} \frac{e^{ik\omega t}}{k} = \sum_k (-1)^{\frac{k-1}{2}} \frac{w_1^k(t)}{ka^k} \qquad k = 1, 3, 5, \ldots \tag{7.40}$$

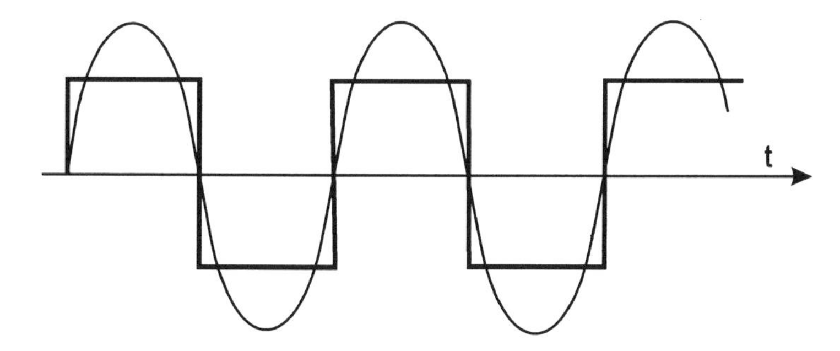

Figure 7.2 Rectangular current in step nonlinearity.

Here, we have taken into account that $w_1 = ae^{i\omega t}$ and have replaced $e^{i\omega t}$ by w_1/a.

Since the current $I(t)$ is known, iterations are needless. For slow modulation $m(\mu t)$, we apply the HT and Bedrosian's theorem to (7.29) to obtain

$$w'' + w = \varepsilon \frac{d}{dt}[w_I(t)m(\mu t) - w]$$

Then, after substitution (7.40) and frequency separation, we find individual equations for the harmonics

$$w_k'' + w_k = \varepsilon \frac{d}{dt}\left[(-1)^{\frac{k-1}{2}} \frac{m}{ka^k} w_1^k - w_k\right] \qquad k = 1, 3, 5, \ldots \tag{7.41}$$

For $k = 1$, this is just the (6.11) with the same function γ:

$$w_1'' + w_1 = \varepsilon \frac{d}{dt}\left[\left(\frac{m}{a} - 1\right) w_1\right]$$

So, concerning the first harmonic, step nonlinearity is completely equivalent to the narrowband generator from Section 6.2, and dynamic frequency fluctuations appear in the same way. Since the rectangular current is independent of higher harmonics, nonlinear interaction and static frequency fluctuations do not appear. Transient regimes are considered in Problem 7.5. They are contrasting for cubic and step nonlinearity.

7.4 Supplementary Problems

Problem 7.1

Using elliptic functions and integrals, find the exact solution of the pendulum equation

$$u'' + \sin u = 0$$

Determine frequency as a function of maximal deviation. Then, using the dependence between maximal deviation and first harmonic, verify the second-order frequency (7.14) obtained with the FSP.

Solution Let us introduce a new variable $p = du/dt$ instead of time t. Then:

$$u'' = \frac{dp}{dt} = \frac{dp}{du}\frac{du}{dt} = p\frac{dp}{du}$$

and the equation takes integrable form:

$$p\frac{dp}{du} = \frac{1}{2}\frac{dp^2}{du} = -\sin u$$

So, we obtain

$$p = \frac{du}{dt} = \sqrt{2(\cos u - \cos A)} = 2\sqrt{\sin^2\frac{A}{2} - \sin^2\frac{u}{2}}$$

Here, A is the maximal deviation when $p = u' = 0$. Integrating again, we find

$$t = \frac{1}{2}\int_0^u \frac{du}{\sqrt{\sin^2\frac{A}{2} - \sin^2\frac{u}{2}}}$$

Then, using new variables ψ and k given by

$$\sin\frac{u}{2} = k\sin\psi \qquad k = \sin\frac{A}{2} \tag{7.42}$$

we come to the elliptic integral $F(\psi, k)$ of the modulus k:

$$t = \int_0^\psi \frac{d\psi}{\sqrt{1 - k^2\sin^2\psi}} = F(\psi, k) \tag{7.43}$$

So, the dependence $t(\psi)$ is given by the elliptic integral $F(\psi, k)$. Jacobian elliptic functions $\mathrm{am}(t, k)$ and $\mathrm{sn}(t, k)$ are defined as follows. The $\mathrm{am}(t, k) = \psi(t)$ is the function inverse to $F(\psi, k)$ and $\mathrm{sn}(t, k) = \sin[\mathrm{am}(t, k)]$. Therefore, (7.42) and (7.43) define exact solution of pendulum equation in the form

$$\sin\frac{u}{2} = k\,\mathrm{sn}(t, k) \tag{7.44}$$

According to (7.42), for $\psi = \pi/2$ and $u = A$, (7.43) defines a quarter of the period:

$$\frac{T}{4} = F\left(\frac{\pi}{2}, k\right) = \frac{\pi}{2}\left(1 + \frac{k^2}{4} + \frac{9k^4}{64} + \cdots\right)$$

where we used the known expansion for $F(\frac{\pi}{2}, k)$. Therefore, as a function of the maximal deviation A, the frequency is given by

$$\omega^2 = \left(\frac{2\pi}{T}\right)^2 = 1 - \frac{k^2}{2} - \frac{3k^4}{32} = 1 - \frac{A^2}{8} + \frac{7A^4}{1536} \tag{7.45}$$

where it was used that $k^2 = \sin^2 \frac{A}{2} = \frac{A^2}{4} - \frac{A^4}{48} + \cdots$.

In Section 7.1, the u was replaced by $\sqrt{\varepsilon}u$, and it was shown that the third harmonic is given by (7.12). For the two harmonics, the maximal deviation is

$$A = \sqrt{\varepsilon}[w_1(0) + w_3(0)] = \sqrt{\varepsilon}a\left(1 - \frac{\varepsilon a^2}{192}\right)$$

Substituting into (7.45), we finally get

$$\omega^2 = 1 - \frac{\varepsilon a^2}{8} + \frac{3\varepsilon^2 a^4}{512}$$

which is just (7.14). So, using the FSP, we obtained exact second-order frequency.

Problem 7.2

Let potential of ion interaction in a crystal be given by Lennard-Johns formula [10]:

$$U = U_0\left[2\left(\frac{r_0}{r}\right)^6 - \left(\frac{r_0}{r}\right)^{12}\right]$$

where r_0 is the equilibrium inter-ion distance. Considering small oscillations of r around r_0, show that the mean inter-ion distance is increasing by

$$\delta r = 5.25\frac{A^2}{r_0} \tag{7.46}$$

where A is amplitude of oscillations. Interpret the increased distance in analogy to the quadratic detector.

Further, define the thermal expansion factor $\alpha = \delta r / r_0 T$ from the well-known relation from kinetic theory

$$\frac{m\omega^2 A^2}{2} = kT \tag{7.47}$$

Here, $k = 1.38 \cdot 10^{-23}$ J/K is the Boltzmannian constant, T is absolute temperature, m is the ion mass, and the left-hand side is the energy of oscillations. How does α vary for other potential functions?

Solution Setting $r = r_0(1 + x)$ for $x \ll 1$, we expand $U(x)$ in a power series and find the backward force as follows:

$$F(x) = -\frac{dU}{dr} = -\frac{1}{r_0}\frac{dU}{dx} = 72\frac{U_0}{r_0}(-x + 10.5x^2 + \cdots) \tag{7.48}$$

Then, the equation of motion $mr_0 x'' = F(x)$ takes the form with the *quadratic* nonlinearity and the nominal frequency $\omega_0^2 = 72U_0/mr_0^2$ (we use nondimensional time $\omega_0 t$ and set $x = \varepsilon u$):

$$u'' + u = 10.5\varepsilon u^2 \tag{7.49}$$

Using $u^2 = (w + w^*)^2/4$, we eliminate negative frequencies and find the first-order equation for the AS:

$$w'' + w = 5.25\varepsilon(w_1^2 + |w_1|^2)$$

Finally, after frequency separation, we obtain the system

$$w_1'' + w_1 = 0$$
$$w_0'' + w_0 = 5.25\varepsilon|w_1|^2$$
$$w_2'' + w_2 = 5.25\varepsilon w_1^2$$

As in the detector, besides the second harmonic, quadratic nonlinearity produces the constant displacement $u_0 = w_0 = 5.25\varepsilon a^2$. In a natural scale, $\delta r = \varepsilon r_0 u_0$, $A = \varepsilon r_0 a$, and we come to (7.46). So, the increased inter-ion distance comes from "detection" in the quadratic nonlinearity.

The thermal expansion factor is as follows:

$$\alpha = \frac{\delta r}{r_0 T} = \frac{10.5k}{mr_0^2\omega_0^2} = \frac{10.5k}{72U_0} = \frac{10.5k}{r_0^3 E}$$

where we have used (7.46), (7.47), and (7.49). Here, $E = 72U_0/r_0^3$ is the modulus of elasticity defined by the linear term in (7.48) for the elementary cell of area r_0^2.

For other potential functions, the factor 10.5 may be changed, but the relation among thermal expansion, elasticity, and inter-ion distance remains the same. Only for very special cases (e.g., molybdenic compounds), does the quadratic term in (7.48) disappear, and thermal expansion is different.

Problem 7.3

In Section 6.2.2 with our study of flicker noise modulation, we restricted the noise band to the generator frequency $\omega_0 = 1$. Out of this band, flicker noise is very small. Nevertheless, for Van der Pol's generator with cubic nonlinearity (7.27), consider *high-frequency* flicker modulation in the band $1 < \omega < 3$. What phenomena come from such modulation?

Solution For $m = 1 + n$, the generator equation takes the form

$$u'' + u = \varepsilon \frac{d}{dt}\left[\left(2u - \frac{4}{3}u^3\right)(1+n) - u\right]$$

and, since $u^{(0)} = u_1$, iteration produces the first-order equation as follows:

$$\frac{d^2u^{(1)}}{dt^2} + u^{(1)} = \varepsilon \frac{d}{dt}\left[\left(2u_1 - \frac{4}{3}u_1^3\right)(1+n) - u_1\right] \tag{7.50}$$

The noise n in the band $1 < \omega < 3$ is concentrated around $\omega = 2$. Its frequency is higher than that of u_1 but lower than that of the third harmonic. Therefore, using Bedrosian's theorem, it can be shown that the addends in the right-hand part are transformed into AS as follows:

$$2u_1(1+n) \Rightarrow 2w_1 + (w_1 + w_1^*)w_n$$

$$\frac{4}{3}u_1^3 \Rightarrow a^2 w_1 + \frac{w_1^3}{6}$$

$$\frac{4}{3}u_1^3 n \Rightarrow \frac{a^2}{2}(w_1 + w_1^*)w_n + \frac{1}{12}w_1^3(w_n + w_n^*)$$

Therefore, the first, third, and fifth harmonics exist in the right-hand side of (7.50). So we set $u^{(1)} = u_1 + u_3 + u_5$ in the left-hand side of (7.50), and collecting the terms at $\omega \approx 1$, come to an equation for the AS w_1 as follows:

$$w_1'' + w_1 = \varepsilon \frac{d}{dt}\left[(1 - a^2)w_1 + \left(1 - \frac{a^2}{2}\right)w_1^* w_n - \frac{1}{12}w_1^3 w_n^*\right]$$

In the right-hand side, the first addend determines the amplitude $a = 1$. The second addend contains w_1^*. Like in (7.18), it produces parametric excitation with the modulating noise around $\omega = 2$. Parametric swinging, however, is now limited because the amplitude is tied to $a = 1$. For $w_1 = e^{it}$ and $w_n = ze^{2it}$, the third addend is $z^* e^{it}$, where z is the complex noise amplitude in the band $-1 < \omega < 1$. Like in (6.20), this *additive* wideband noise around $\omega = 1$ produces amplitude and frequency distortions. So, as assumed in Section 6.2.2, only *low-frequency* flicker noise at $0 < \omega < 1$ produces modulation. Its *high-frequency* components produce different effects—parametric excitation and additive noise distortions similar to, but much less than, thermal and shot noises.

Problem 7.4

In Section 6.2.3, we restricted the additive noise to the double generator frequency $\omega = 2$. Reassert this approach with frequency separation.

Solution In a noise-free case, the first-order oscillation consists of the first and third harmonics, and we separate them at any point between $\omega = 1$ and $\omega = 3$. In a noisy background, separating the harmonics, we also demarcate the noise zones around $\omega = 1$ and $\omega = 3$. Now, however, the bound should be chosen at the middle $\omega = 2$. Then, the noise at $0 < \omega < 2$ distorts the first harmonic, whereas the noise at $2 < \omega < 4$ distorts the third harmonic. Interaction among the noises is negligible for the first order, and this approach is allowable in Section 6.2.3.

Problem 7.5

According to (6.12) and (6.13), amplitude and frequency of a generator obey equations (we replace K by a^2):

$$a' = \frac{\varepsilon}{2}\gamma(a)a \qquad \omega^2 = 1 - \frac{a''}{a} \tag{7.51}$$

In Chapter 6, these equations were derived for narrowband generators, but in Section 7.3, we also showed that they are right for wideband generators if the function $\gamma(a)$ is defined by (we set $m = 1$):

$$\gamma(a) = \begin{cases} 1 - a^2 & \text{for cubic nonlinearity} \\ \dfrac{1}{a} - 1 & \text{for step nonlinearity} \end{cases} \tag{7.52}$$

After switching on, the generator amplitude and frequency are varying. Find transient characteristics describing these variations.

Solution Amplitude equation (7.51) takes the form

$$a' = \begin{cases} \frac{\varepsilon}{2}(a - a^3) & \text{for cubic nonlinearity} \\ \frac{\varepsilon}{2}(1 - a) & \text{for step nonlinearity} \end{cases} \tag{7.53}$$

The second equation is linear, and the first one is a Bernoulli equation. Solving them (see Problem 6.7), we obtain

$$a(t) = \begin{cases} \dfrac{a_0}{\sqrt{a_0^2 + (1 - a_0^2)\, e^{-\varepsilon t}}} & \text{for cubic nonlinearity} \\ 1 - (1 - a_0)e^{-\varepsilon t/2} & \text{for step nonlinearity} \end{cases} \tag{7.54}$$

where a_0 is the initial amplitude at $t = 0$. These relations define the transient amplitudes shown in Figure 7.3.

Differentiating the first equation (7.51) and substituting into the second, we also obtain transient frequencies in the form

$$\omega(t) = \sqrt{1 - \frac{\varepsilon^2}{4}[\gamma'(a)a + \gamma(a)]\gamma(a)}$$

For cubic nonlinearity, (7.38) defines the additional static correction $\Delta\omega = -\frac{\varepsilon^2}{16}a^4$. For the amplitudes (7.54), transient frequencies are also shown in Figure 7.3. Clearly, transient characteristics are nonlinearity-dependent, and dynamic distortions dominate.

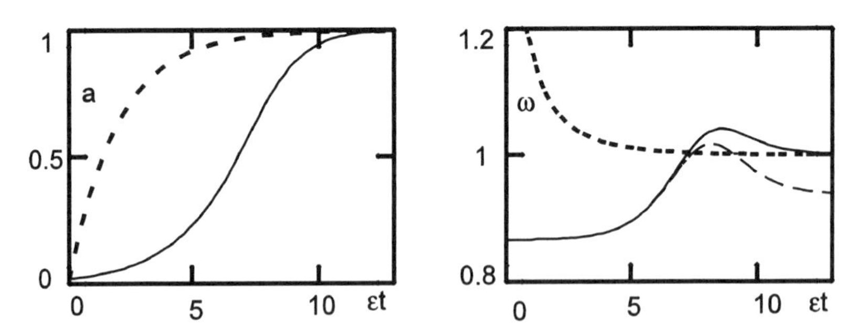

Figure 7.3 Amplitude and frequency transient characteristics for cubic (solid lines) and step (dotted lines) nonlinearity. Dashed line shows additional static frequency correction for cubic nonlinearity. Frequency deviations are depicted for $\varepsilon = 1$; in reality, they are multiplied by ε^2.

For cubic nonlinearity, the amplitude and static frequency characteristics were found by Van der Pol in the 1930s [1]. However, neither transient amplitudes for step nonlinearity nor dynamic frequency characteristics have been found until now.

References

[1] Van der Pol, B., "The Nonlinear Theory of Electrical Oscillations," *Proc. of the IRE*, Vol. 22, 1934, pp. 1051–1086.

[2] Bogoliuboff, N. N., and Y. A. Mitropolskii, *Asymptotic Methods in the Theory of Nonlinear Oscillations*, New York: Gordon and Breach, 1961.

[3] Hayashi, C., *Nonlinear Oscillation in Physical Systems*, New York: McGraw-Hill, 1964.

[4] Minorsky, N., *Nonlinear Oscillations*, Princeton, NJ: D. Van Nostrand, 1962.

[5] Wainstein, L. A., and D. E. Vakman, *Frequency Separation in Theory of Oscillations and Waves*, Moscow: Nauka-Press, 1983 (in Russian).

[6] Gilmore, R. J., and M. B. Steer, "Nonlinear Circuit Analysis Using the Method Harmonic Balance—A Review of the Art," *Int. J. Microwave Millimeter-Wave Computer-Aided Eng.*, Vol. 1, 1991, pp. 22–37 and 159–180.

[7] Kevorkian, J., and J. D. Cole, *Perturbation Methods in Applied Mathematics*, New York: Springer-Verlag, 1981.

[8] Nayfeh, A. H., and D. T. Mook, *Nonlinear Oscillations*, New York: Wiley, 1979.

[9] Buonomo, A., and C. D. Bello, "Asymptotic Formulas in Nearly Sinusoidal Nonlinear Oscillators," *IEEE Trans. Circuits and Systems*, pt. 1, Vol. 43, 1996, pp. 953–963.

[10] Livshitz, B. G., *Physical Characteristics of Metals and Alloys*, Moscow: Mashgiz, 1956 (in Russian).

8

Electrons in Crossed Fields

In Section 6.3.2, we have used an equation of electron motion in the crossed electric and magnetic fields:

$$w'' = \frac{e}{m}E - i\omega_c w' \qquad \omega_c = \frac{eH}{mc} \tag{8.1}$$

Here, $w = u + iv$ is the complex coordinate of the electron (not AS), $E = E_u + iE_v$ is the complex electric field, ω_c is the cyclotron frequency, and $-i\omega_c w'$ is the velocity-dependent magnetic force. We have studied a rotating electric field that excites a resonant motion. Now, a *static* electric field is considered that obeys the Laplacian equation with the potential function $U(u, v)$, such that

$$E_u = \frac{\partial U}{\partial u} \qquad E_v = \frac{\partial U}{\partial v} \quad \text{where } \frac{\partial^2 U}{\partial u^2} + \frac{\partial^2 U}{\partial v^2} = 0 \tag{8.2}$$

For a spatial charge, we will also consider the Poissonian field.

The electron motion is usually studied with averaging [1–3]. In the magnetic field H, the electron rotates on a circular orbit, while the orbit drifts along the equipotential line $U(u, v) = \text{const}$ in the electric field. For magnetrons, the motion is commonly studied in the moving coordinate system, where the running wave in the anode-cathode space becomes static (see also Problem 8.5).

Fast orbital rotation and slow drift are the motions of separable frequencies (the drift is at zero frequency). Using the FSP developed in Chapter 7, we investigate the motion up to the fourth order and find out the following additional effects:

- The orbit becomes elliptical and its size varies from point to point.
- Orbital velocity also varies so that energy of rotation is conserved.
- Additional epicycles appear with fast-backward rotations: the main orbit is a trajectory of the center of a smaller orbit, and so on.
- In a nonuniform electric field, the drift trajectory undergoes higher-order perturbations and deviates from the equipotential line.

As in Chapter 7, these effects come from the nonlinear interaction of components with different frequencies. We emphasize again that the electron motion is not the AS, and its components are at positive and negative frequencies. However, each narrowband component can be interpreted as the AS.

8.1 First-Order Solution

8.1.1 Basic Equation

In (8.1), the complex conjugate coordinates $w = u + iv$ and $w^* = u - iv$ are used instead of the real coordinates u and v, and generally the field $E = E(w, w^*)$ depends on both w and w^*. However, since $u = (w + w^*)/2$ and $v = (w - w^*)/2i$, we have

$$\frac{\partial E}{\partial w} = \frac{\partial E}{\partial u}\frac{\partial u}{\partial w} + \frac{\partial E}{\partial v}\frac{\partial v}{\partial w}$$

$$= \frac{\partial(E_u + iE_v)}{\partial u} \cdot \frac{1}{2} + \frac{\partial(E_u + iE_v)}{\partial v} \cdot \frac{1}{2i}$$

$$= \frac{1}{2}\left(\frac{\partial^2 U}{\partial u^2} + \frac{\partial^2 U}{\partial v^2}\right) + \frac{i}{2}\left(\frac{\partial E_v}{\partial u} - \frac{\partial E_u}{\partial v}\right) = 0 \qquad (8.3)$$

Here, we have taken into account the Laplacian equation (8.2) and that $\partial E_v/\partial u = \partial E_u/\partial v = \partial^2 U/\partial u\, \partial v$. So, the *static* electric field E depends on w^* only, and (8.1) can be rewritten as follows:

$$w'' + iw' = \varepsilon f(w^*) \quad \text{where } \varepsilon = \frac{eE_0}{m\omega_c^2 D} = \frac{mc^2 E_0}{eH^2 D} \ll 1 \qquad (8.4)$$

if the electric field is weak. Here we have introduced the nondimensional time $\omega_c t$ and the nondimensional coordinate w/D, where D is a typical size like anode-cathode distance. Nevertheless, we save the notations t and w for the nondimensional time and coordinate. We have also introduced the maximal (or mean) electric field E_0 and denoted $E = E_0 f(w^*)$.

8.1.2 Preliminary Examples

Let the electric field be zero and $f(w^*) = 0$. Then, (8.4) can easily be solved and we find

$$w = w_0 + iv_0(e^{-it} - 1)$$

Here, w_0 and v_0 are the initial position and velocity at $t = 0$. So, the electron runs on a circular orbit with the angular velocity $\omega = -1$. The orbital radius depends on the initial velocity: the nondimensional radius is $R = v_0$, whereas natural values are $\omega = -\omega_c$ and $R = v_0/\omega_c = mcv_0/eH$. Obviously, the rotation comes from the magnetic force orthogonal to velocity at each point.

A more complex example is a constant electric field $f = 1$, but the linear equation (8.4) is also solvable, and we find

$$w = w_0 - i\varepsilon t + (iv_0 - \varepsilon)(e^{-it} - 1)$$

As before, the electron rotates, but in addition, its orbit drifts with a small constant speed, $-i\varepsilon$. Because of the factor i, the drift is perpendicular to the field $f = 1$ and occurs along the equipotential line.

For this example, the drift frequency is $\omega = 0$ while the rotation frequency is $\omega = -1$. For a nonuniform electric field, their spectra are wider, with $\Delta\omega \sim \varepsilon$, and additional higher harmonics appear. So, the general solution of (8.4) takes the form

$$w(t) = \sum_k w_k(t) = \sum_k Z_k(\varepsilon t)e^{ikt} \qquad k = 0, \pm 1, \pm 2, \ldots \tag{8.5}$$

where complex amplitudes Z_k depend on slow time εt, and the $w_0(t) = Z_0(\varepsilon t)$ describes the drift.

8.1.3 Iterations and Frequency Separation

Now, for the electric field $f(z^*)$, we apply the FSP to (8.4) assuming a *small* initial velocity $v_0 \sim \varepsilon$ and a small orbit (for large v_0, see Problems 8.3 and 8.4). Then, the drift is the zeroth-order approximation, $w^{(0)} = w_0 = Z_0(\varepsilon t)$, and the first-order iterative equation (7.7) takes the form

$$\frac{d^2w^{(1)}}{dt^2} + i\frac{dw^{(1)}}{dt} = \varepsilon f(w_0^*)$$

As we have seen, its solution consists of the slow drift w_0 and fast rotation w_{-1}. Therefore, we set $w^{(1)} = w_0 + w_{-1}$ and, separating frequencies, obtain two equations:

$$\begin{aligned} w_0'' + iw_0' &= \varepsilon f\left(w_0^*\right) \\ w_{-1}'' + iw_{-1}' &= 0 \end{aligned} \tag{8.6}$$

The first equation defines the drift. Formally, this is the same equation as (8.4), but since the drift depends on εt, its second derivative w_0'' is of the order ε^2. Neglecting it in the first order, we find

$$iw_0' = \varepsilon f\left(w_0^*\right) \tag{8.7}$$

This equation shows that the electron drifts along the equipotential line $U = const.$ Indeed, multiplying (8.7) by $w_0'^*$ and taking the real part, we find

$$\mathrm{Re}\left[f\left(w_0^*\right)w_0'^*\right] = \mathrm{Re}\left[2\frac{dU(w_0^*)}{dw_0^*}w_0'^*\right] = 2\frac{dU}{dt} = 0$$

and therefore, potential is constant in the drift trajectory. Here we have used that $f = f_u + if_v = \partial U/\partial u + i\partial U/\partial v = 2\partial U/\partial w^*$ and that potential $U(u, v)$ is a real function.

The second equation (8.6) defines the orbital rotation

$$w_{-1} = Z_{-1}e^{-it} \quad \text{where } Z_{-1} \sim v_0 \sim \varepsilon \tag{8.8}$$

for a small initial velocity. Equations (8.7) and (8.8) give the first-order solution and describe basic electron motions. They are often found with averaging.

8.2 Higher-Order Solutions

8.2.1 Second Order

Now setting $w^{(1)} = w_0 + w_{-1}$, where $w_{-1} = Z_{-1}e^{-it} \sim \varepsilon$, we apply the iterative procedure again. Then the second-order equation takes the form:

$$\frac{d^2w^{(2)}}{dt^2} + i\frac{dw^{(2)}}{dt} = \varepsilon f(w_0^* + Z_{-1}^*e^{it}) = \varepsilon f\left(w_0^*\right) + \varepsilon f'\left(w_0^*\right)Z_{-1}^*e^{it}$$

where in the right-hand side, we kept the addends up to the second order.

The addend with e^{it} in the right-hand side is a backward rotation with $\omega = 1$. Therefore, $w^{(2)} = w_0 + w_{-1} + w_1$, and after frequency separation, we come to the second-order system as follows:

$$\begin{aligned} w_0'' + i w_0' &= \varepsilon f(w_0^*) \\ w_{-1}'' + i w_{-1}' &= 0 \\ w_1'' + i w_1' &= \varepsilon f'(w_0^*) w_{-1}^* \end{aligned} \tag{8.9}$$

As before, two equations show the slow drift w_0 and basic rotation w_{-1}. However, the third equation shows a new effect—the backward rotation with the frequency $\omega = 1$. As usual, when solving it, we keep the forced motion and neglect derivatives of slow factors. Then we find

$$w_1 = -\frac{\varepsilon}{2} f'(w_0^*) w_{-1}^* \tag{8.10}$$

A sum of forward and backward rotations,

$$w_{-1} + w_1 = Z_{-1} e^{-it} - \frac{\varepsilon}{2} f'(w_0^*) Z_{-1}^* e^{it}$$

results in the elliptical orbit with a small eccentricity of the order ε.

8.2.2 Third Order

Substituting the second-order solution $w^{(2)} = w_0 + w_{-1} + w_1$ into the right-hand side of (8.4) and separating frequencies, we obtain equations of the third order:

$$\begin{aligned} w_0'' + i w_0' &= \varepsilon f(w_0^*) \\ w_{-1}'' + i w_{-1}' &= \varepsilon f'(w_0^*) w_1^* \\ w_1'' + i w_1' &= \varepsilon f'(w_0^*) w_{-1}^* \\ w_2'' + i w_2' &= \frac{\varepsilon}{2} f''(w_0^*) w_{-1}^{*2} \end{aligned} \tag{8.11}$$

The right-hand side in the equation for w_{-1} varies the basic rotation. When substituting into this equation, the function w_1 is to be found up to the second order ε^2, and for this accuracy, solution of the third equation is the same as in

(8.10). Therefore, the equation for w_{-1} takes the form

$$w''_{-1} + iw'_{-1} + \frac{\varepsilon^2}{2}|f'(w_0^*)|^2 w_{-1} = 0$$

Using Problems 6.3 and 6.2, we obtain its solution as follows:

$$w_{-1} = \frac{\text{const}}{\sqrt{1+\varepsilon^2|f'(w_0^*)|^2}} \exp -i\left(t + \frac{\varepsilon^2}{2}\int |f'(w_0^*)|^2\, dt\right) \tag{8.12}$$

So, when drifting, the orbital radius and angular velocity are varied so that energy of rotation remains constant: $|z_{-1}(t)|\omega(t) = const$ (see general approach in Problem 8.2).

The last equation (8.11) defines an additional backward rotation of the double frequency $\omega = 2$:

$$w_2 = -\frac{\varepsilon}{12} f''(w_0^*) w_{-1}^{*2} = -\frac{\varepsilon}{12} f''(w_0^*)\, Z_{-1}^{*2} e^{i2t} \tag{8.13}$$

For $Z_{-1} \sim \varepsilon$, its amplitude (radius of rotation) is of the order ε^3. This fast rotation can be interpreted as a small epicycle, and the basic orbit is a trajectory of its center.

8.2.3 Fourth Order

In the same way, substituting $w^{(3)} = w_0 + w_{-1} + w_1 + w_2$ into the right-hand side of (8.4), after frequency separation we obtain the fourth-order system:

$$\begin{aligned}
w_0'' + iw_0' &= \varepsilon f(w_0^*) + \varepsilon f''(w_0^*) w_{-1}^* w_1^* \\
w''_{-1} + iw'_{-1} &= \varepsilon f'(w_0^*) w_1^* \\
w_1'' + iw_1' &= \varepsilon f'(w_0^*) w_{-1}^* \\
w_2'' + iw_2' &= \frac{\varepsilon}{2} f''(w_0^*) w_{-1}^{*2} \\
w''_{-2} + iw'_{-2} &= \varepsilon f'(w_0^*) w_2^* \\
w_3'' + w_3' &= \frac{\varepsilon}{6} f'''(w_0^*) w_{-1}^{*3}
\end{aligned} \tag{8.14}$$

This system shows several new effects. First, for a nonuniform field with $f''(w_0^*) \neq 0$, the equation for the drift w_0 contains a small correction of the order ε^4 (we remind that $w_{-1} \sim \varepsilon$ and $w_1 \sim \varepsilon^2$). Like in the synchronous detector, forward and backward rotations w_{-1} and w_1 result in a slow component that distorts the drift motion. This can also be explained in another way. In a nonuniform electric field, when rotating on the elliptic orbit, a constant force (averaged over the orbit) produces an additional slow motion.

Further, besides the backward rotation w_2 with $\omega = 2$, the smaller forward rotation w_{-2} also appears, and the second epicycle becomes elliptical. Finally, the third epicycle w_3 of the smallest size ε^4 appears at the triple frequency $\omega = 3$.

Clearly, further higher-order solutions result in the smaller epicycles. So, the classical electron motion consisting of the drift and circular rotation is an approximation. In reality, additional elliptic epicycles appear with fast-backward rotations, and the drift trajectory deviates from the equipotential line.

8.3 Spatial Charge Effects

Now, instead of the electrostatic Laplacian field, we consider a Poissonian field with a spatial charge density $\rho(u, v)$. We take the Poissonian field equation in the form

$$\frac{\partial^2 U}{\partial u^2} + \frac{\partial^2 U}{\partial v^2} = \frac{\omega_p^2}{\varepsilon \omega_c^2}$$

Here $\omega_p = \sqrt{4\pi e\rho/m}$ is the *plasmic frequency*, and the field $f(w^*, w)$ depends on both w^* and w. Its derivatives are as follows:

$$\begin{aligned} f_1 &= \frac{\partial f}{\partial w^*} = \frac{1}{2}\left(\frac{\partial^2 U}{\partial u^2} - \frac{\partial^2 U}{\partial v^2}\right) - i\frac{\partial^2 U}{\partial u\, \partial v} \\ f_2 &= \frac{\partial f}{\partial w} = \frac{1}{2}\left(\frac{\partial^2 U}{\partial u^2} + \frac{\partial^2 U}{\partial v^2}\right) = \frac{\omega_p^2}{2\varepsilon \omega_c^2} \end{aligned} \tag{8.15}$$

and the dependence on w comes from the spatial charge. If the density ρ becomes critical ($\omega_p \sim \omega_c$), the functions εf_2 and εf are *not small*, and our method fails. However, for a small charge density ($\omega_p^2/\omega_c^2 \sim \varepsilon$), the method remains valid, and the basic equation remains the same as in (8.4):

$$w'' + iw' = \varepsilon f(w^*, w) \tag{8.16}$$

The first-order solution is also the same: the electron rotates on a circular orbit and drifts along the equipotential line (of the field distorted with a spatial charge).

Distinctions appear in higher orders. For the second order, substituting $w^{(1)} = w_0 + w_{-1}$ into the right-hand side of (8.16), we obtain the iterative equation

$$\begin{aligned}\frac{d^2 w^{(2)}}{dt^2} + i\frac{dw^{(2)}}{dt} &= \varepsilon f(w_0^* + w_{-1}^*,\ w_0 + w_{-1}) \\ &= \varepsilon f\,(w_0^*, w_0) + \varepsilon f_1\,(w_0^*, w_0) w_{-1}^* + \varepsilon f_2\,(w_0^*, w_0) w_{-1}\end{aligned}$$

where f_1 and f_2 are the derivatives (8.15). Here the right-hand side contains three frequencies: $\omega = 0$, $\omega = -1$, and $\omega = 1$. Therefore, $w^{(2)} = w_0 + w_{-1} + w_1$, and after frequency separation, we find three equations:

$$\begin{aligned} w_0'' + iw_0' &= \varepsilon f \\ w_{-1}'' + iw_{-1}' &= \varepsilon f_2 w_{-1} \\ w_1'' + iw_1' &= \varepsilon f_1 w_{-1}^* \end{aligned} \tag{8.17}$$

This second-order system is similar to the *third-order* system (8.11). The equation for w_{-1} contains the right-hand side, and its solution is analogous to (8.12):

$$w_{-1} = \frac{\text{const}}{\sqrt{1 - \omega_p^2/\omega_c^2}} \exp -i\left(t - \frac{1}{2\omega_c^2}\int \omega_p^2\, dt\right) \tag{8.18}$$

So, a spatial charge reduces the frequency of rotation. The orbital radius, however, increases in such a degree that energy of rotation is conserved. In contrast to a Laplacian field, these effects appear in the second order.

For the third order, substituting $w^{(2)} = w_0 + w_{-1} + w_1$ into the right-hand side of (8.16), we obtain the system analogous to (8.14). It defines a correction to the drift motion. So, a spatial charge shifts the above effects by one order: orbital ellipticity, epicycles, and drift distortions arise earlier and in greater extent.

8.4 Supplementary Problems

Problem 8.1

Using (8.4), reassert the well-known fact that electron energy is independent of the magnetic field. Show also that total electron energy is conserved.

Solution Multiplying (8.4) by w'^* and taking the real part, we remove the magnetic force and find

$$\mathrm{Re}[w''w'^*] = \varepsilon\,\mathrm{Re}[f(w^*)w'^*]$$

Further, since $2\,\mathrm{Re}[w''w'^*] = \frac{d}{dt}|w'|^2$ and $\mathrm{Re}[f(w^*)w'^*] = \frac{dU}{dt}$, integrating in time, we find the conservation law as follows:

$$\frac{1}{2}|w'|^2 - \varepsilon U = \mathrm{const} \tag{8.19}$$

Here, $|w'|^2/2$ is the kinetic energy and $-\varepsilon U$ is the potential energy of the electron. Whatever the magnetic field, their sum is conserved.

Problem 8.2

As shown in Section 8.1, the drift w_0 obeys (8.7) and occurs along the equipotential line. This is true, however, for the first order only. The more precise drift equation [up to the third order—compare (8.6), (8.9), and (8.11)] is as follows:

$$w_0'' + iw_0' = \varepsilon f(w_0^*) \tag{8.20}$$

Show that drift energy is conserved. Consider conservation laws for other partial motions (basic rotation, the first epicycle, etc.) Show that conservation of partial energies comes from the frequency separability.

Solution Formally, (8.20) is the same as (8.4), and according to Problem 8.1, the drift energy is conserved [potential $U(w_0)$ is now taken for the leading center w_0]. Equations (8.12) and (8.18) also show that energy of basic rotation is conserved, and the same applies to other rotations.

Physically, this comes from the fact that the components of motion are of separable spectral bands, and their spectra are nonoverlapping. Therefore, energies of the drift and rotations do not interchange one another, and since the total energy is conserved, each partial energy is conserved also.

Problem 8.3

In the first order, (8.4) defines the orbital rotation dependent on the initial velocity v_0, and we have assumed a small velocity $v_0 \sim \varepsilon$. What is the solution for a large initial velocity $v_0 \sim 1$ and a large orbital radius $w_{-1} \sim 1$?

Solution For $w_{-1} \sim 1$, the first-order equation contains all terms of the Taylor series:

$$\frac{d^2 w^{(1)}}{dt^2} + i\frac{dw^{(1)}}{dt} = \varepsilon f\left(w_0^* + w_{-1}^*\right)$$
$$= \varepsilon f\left(w_0^*\right) + \varepsilon f'(w_0^*)w_{-1}^* + \frac{\varepsilon}{2} f''(w_0^*)w_{-1}^{*2} + \cdots$$

Therefore, after frequency separation, we come to the infinite system for all higher harmonics:

$$w_0'' + iw_0' = \varepsilon f\left(w_0^*\right)$$
$$w_{-1}'' + iw_{-1}' = 0$$
$$w_k'' + iw_k' = \frac{\varepsilon}{k!} f^{(k)}\left(w_0^*\right)w_{-1}^{*k} \qquad k = 1, 2, 3, \ldots$$

Equations for the drift and basic rotation are the same as in (8.6). Even for the first order however, all backward rotations (epicycles) appear:

$$w_k = -\frac{\varepsilon}{(k+1)!k} f^{(k)}\left(w_0^*\right)w_{-1}^{*k}$$

and the main orbit becomes elliptical.

The second-order equation takes the form

$$\frac{d^2 w^{(2)}}{dt^2} + i\frac{dw^{(2)}}{dt} = \varepsilon f\left(w_0^* + w_{-1}^* + w_1^* + w_2^* + \cdots\right)$$

and nonlinear interaction of all harmonics should be considered. This results in forward rotations, and all epicycles become elliptical.

Another approach is also effective. The drift w_0 is of secular behavior, whereas orbital motion, even not small, is bounded. The convergent series above can be interpreted as formal expansions with respect to a new parameter $|w_{-1}| = \varepsilon_1$. Then, formal relations from Section 8.2 remain valid.

Problem 8.4

Setting $f = f(\varepsilon w^*)$, simplify the previous problem for weak nonuniformity of the field.

Solution If the field depends on εw^*, the first-order and second-order iterative equations are the same as for $w_{-1} \sim \varepsilon$:

$$\frac{d^2 w^{(1)}}{dt^2} + i\frac{dw^{(1)}}{dt} = \varepsilon f\left(\varepsilon w_0^*\right)$$
$$\frac{d^2 w^{(2)}}{dt^2} + i\frac{dw^{(2)}}{dt} = \varepsilon f\left(\varepsilon w_0^*\right) + \varepsilon^2 f'\left(\varepsilon w_0^*\right) w_{-1}^*$$

Therefore, weak nonuniformity results in the same answer as for a small initial velocity.

Problem 8.5

In the plane magnetron, electrons move in the field of a high-frequency running wave $f(w^* - Vt)$, where V is its phase velocity. Besides, the drift velocity V_0 is constant due to the strong uniform electrostatic field (we suppose both V and V_0 to be real, so that the two motions are along the u-axis). If the coordinate system also moves with the drift velocity V_0, equation of motion takes the form

$$w'' + iw' = \varepsilon f(w^* - vt)$$

where $v = V - V_0$ is the relative field velocity.

Explain orbital resonances in magnetrons for the harmonic wave $f(w^*) = \cos w^*$ and for small and large initial velocities.

Solution For a small initial velocity ($w_{-1} \sim \varepsilon$), the first-order equation takes the form

$$\frac{d^2 w^{(1)}}{dt^2} + i\frac{dw^{(1)}}{dt} = \varepsilon \cos\left(w_0^* - vt\right) = \frac{\varepsilon}{2}\left[e^{i(w_0^* - vt)} + e^{-i(w_0^* - vt)}\right]$$

A resonance occurs if $v = \pm 1$ (more precisely, $|v \mp 1| < \varepsilon$) when the right-hand side contains the oscillation at $\omega = -1$. For $v = -1$, frequency separation results in three equations:

$$w_0'' + iw_0' = 0$$

$$w_{-1}'' + iw_{-1}' = \frac{\varepsilon}{2} e^{-i(w_0^* + t)}$$

$$w_1'' + iw_1' = \frac{\varepsilon}{2} e^{i(w_0^* + t)}$$

Here, the first equation gives $w_0' = 0$ and $w_0 = \text{const}$. Therefore, in the motionless coordinates, the drift velocity is V_0. The second equation defines a resonant rotation on the growing orbit

$$w_{-1} = \frac{i\varepsilon t}{2} e^{-i(w_0^* + t)}$$

and the third equation results in a small backward rotation

$$w_1 = -\frac{\varepsilon}{4} e^{i(w_0^* + t)}$$

For $v \neq -1$, in the second order we have

$$\frac{d^2 w^{(2)}}{dt^2} + i\frac{dw^{(2)}}{dt} = \varepsilon \cos\left(w_0^* - vt\right) - \varepsilon(w_+ + w_-) \sin\left(w_0^* - vt\right)$$

where $w_\pm$ are small nonresonant rotations at $\omega = \pm v \neq -1$. However, for $v = \pm 1/2$, the product $w_\pm \cdot \sin(w_0^* - vt)$ contains the frequency $\omega = -1$, and the second-order resonance occurs. In higher orders, the resonances also occur at $v = 1/3$, $v = 1/4$, and so on.

For a large initial velocity ($w_{-1} \sim 1$), even in the first order, the right-hand side contains the term

$$e^{\pm i(w_0^* + w_{-1}^* - vt)} = \exp \pm i\left(w_0^* + Z_{-1}^* e^{it} - vt\right)$$
$$= e^{\pm i w_0^*} \cdot \sum_{n=0}^{\infty} \frac{\left(\pm i Z_{-1}^*\right)^n}{n!} e^{i(n \pm v)t}$$

and resonances also occur at $v = \pm 2, \ \pm 3, \ldots$.

The physical cause of resonances is the same as in Section 6.3.2. At a certain asynchronous drift (with respect to the running wave), the rotating field excites resonant oscillations. Since we have neglected relativistic mass variations, the orbit grows infinitely. In reality, the increasing orbit becomes decreasing, and the resonant motions are bounded.

Problem 8.6

Consider additional orbital resonances coming from spatial harmonics of the field

$$f(w^*) = \sum_k f_k e^{ikw^*}$$

Solution For a small initial velocity, the first-order equation takes the form

$$\frac{d^2 w^{(1)}}{dt^2} + i\frac{dw^{(1)}}{dt} = \varepsilon \sum_k f_k e^{ik(w_0^* - vt)}$$

and resonances appear at $kv = 1$; that is, at $v = \pm 1, \ \pm 1/2, \ \pm 1/3, \ldots$. This is similar to the higher-order resonances for a single spatial harmonic.

References

[1] Bogoliuboff, N. N., and Y. A. Mitropolskii, *Asymptotic Methods in the Theory of Nonlinear Oscillations*, New York: Gordon and Breach, 1961.

[2] Kapitza, P. L., *High Power Electronics*, Moscow: Nauka-Press, 1962 (in Russian).

[3] Wainstein, L. A., and V. A. Solntzev, Lectures on Ultra-High-Frequency Electronics, Moscow: Sov. Radio, 1973 (in Russian).

9

Analytic Waves

Analytic waves (AW) generalize the AS and neatly define the amplitudes and frequencies of real propagating waves [1]. Using the AW, we justify and clarify some points of wave theory related to propagation in nonuniform dispersive media. In Section 9.1, after a brief review of parent relations, we introduce the AW and running spectra of waves. In Section 9.2, we study the center and duration of a wave. It is shown that the local group delay averaged in frequency defines velocity of the wave center at each point. In Section 9.3, asymptotic solution is developed for running waves. Also, Whitham's method [2] is modified for not only frequency but also amplitude of a wave.

Quantum mechanical wave packets are considered in Section 9.4. A wave packet is often the AS, and for a particle moving in one direction, it is also the AW. This provides a physical basis for the AS and AW. Besides, the center of a wave packet represents the associated classical particle, and the paradox of tunneling is clarified. For a particle tunneling through a barrier, the paradox is concerned with a transmission time. A number of approaches have recently been suggested for defining this time [3,4], but time of tunneling is still an open question. However, the paradox disappears if we associate the wave packet with an ensemble of particles instead of a single particle. This paradox is not specifically quantum but occurs and can be explained in a classical area.

In Section 9.5, generalizing the *frequency separation procedure* (FSP) developed for nonlinear oscillations, we study nonlinear waves. *Solitary waves* [5] propagate in nonlinear dispersive media without distortions. This amazing wave phenomenon can easily be considered with the AW.

9.1 Parent Relations

One-dimensional scalar waves in nonuniform dispersive media obey the equation

$$\frac{\partial^2 u}{\partial x^2} - L\left(x, \frac{\partial}{\partial t}\right)u = 0 \quad \text{where } L\left(x, \frac{\partial}{\partial t}\right) = \sum_n \alpha_n(x)\frac{\partial^n}{\partial t^n} \tag{9.1}$$

Here, $u(x, t)$ is a real field (wave) at a point x at an instant t. Our aim is to explore the amplitude and frequency of propagating waves.

The linear differential operator L, as given in (9.1), defines the dispersion properties of a medium. Writing $u(x, t)$ in the form

$$u(x, t) = \frac{1}{(2\pi)^2}\int_{-\infty}^{\infty}\int_{-\infty}^{\infty} U(k, \omega)e^{-i(\omega t - kx)}\, d\omega\, dk \tag{9.2}$$

where $U(k, \omega)$ is the two-dimensional Fourier transform of $u(x, t)$, and substituting (9.2) into (9.1), we obtain

$$\frac{1}{(2\pi)^2}\int_{-\infty}^{\infty}\int_{-\infty}^{\infty} U(k, \omega)[-k^2 - L(x, -i\omega)]e^{-i(\omega t - kx)}\, d\omega\, dk = 0$$

Exponential functions $e^{-i(\omega t - kx)}$ are orthogonal for unequal ω or k. Therefore, the bracket has to be zero, and we come to the wave number k as a function of ω and x:

$$k^2(x, \omega) = -L(x, -i\omega) \quad \text{or} \quad k(x, \omega) = \pm\sqrt{-L(x, -i\omega)} \tag{9.3}$$

Here, $\pm$ specifies the direction of propagation. If L is independent of x, the medium is uniform and k depends on ω only. If also

$$L = \frac{1}{c^2}\frac{\partial^2}{\partial t^2} \quad \text{and} \quad k = \pm\frac{\omega}{c}$$

no dispersion exists, and a signal propagates without distortions with a constant speed c.

Many wave equations can be reduced to (9.1). For the beam equation

$$\frac{\partial^4 u}{\partial x^4} + \alpha^2(x)\frac{\partial^2 u}{\partial t^2} = 0$$

the dispersion relation takes the form $k^4 = \alpha^2(x)\omega^2$. Then we have $k^2 = \pm\alpha(x)\omega$, and using the correspondence $\partial/\partial x \leftrightarrow ik$ and $\partial/\partial t \leftrightarrow -i\omega$, we come to (9.1) with $L = \pm i\alpha(x)\partial/\partial t$.

In turn, the wave number determines the group delay $\tau(x, \omega)$ and the group speed $v(x, \omega)$:

$$\tau(x, \omega) = \frac{\partial k}{\partial \omega} \qquad v(x, \omega) = \frac{1}{\tau} = \frac{\partial \omega}{\partial k} \tag{9.4}$$

In a strict sense, the group speed and delay are defined for harmonic waves or harmonic components of a wave. They are also, however, used generally when replacing the spectral frequency ω by the local (instantaneous) frequency $\omega(x, t)$ that relates to the global and local frequency notions. In particular, Whitham's method discussed in Section 9.3.2 defines the local frequency and local wave number as functions of t and x.

Analytic Waves The amplitude $a(x, t)$, phase $\phi(x, t)$, and instantaneous frequency $\omega(x, t) = \partial\phi/\partial t$ of a wave are often introduced by

$$u(x, t) = a(x, t) \cos \phi(x, t) \tag{9.5}$$

For the same reason as for signals, however, they are ambiguous within a framework of real waves.

To define the amplitude and frequency unambiguously, we introduce the *complex* wave $w(x, t)$, which is the AS as a function of either t or x. This complex wave is referred to as the *analytic wave.* As in (2.12), for a real wave given in (9.2), the associated AW is

$$w(x, t) = a(x, t)e^{i\phi(x,t)} = \frac{1}{\pi^2} \int_0^\infty \int_0^\infty U(k, \omega)e^{-i(\omega t - kx)}\, dk\, d\omega \tag{9.6}$$

In time and space, the amplitudes and frequencies of the AW have the same properties as the AS. The functions of type (9.6) have been studied in [6].

The Fourier transform (9.2) differs from the standard form for signals (2.9) in the sign of ω. This distinction is traditional for waves, and for the form (9.2), the AS w should be replaced by its complex conjugate $w^* = u - iv$ with a spectrum at negative frequencies. However, the amplitude and frequency of real waves remain the same, and we ignore this distinction in (9.6) correlating any *one-sided* spectrum (at positive or negative frequencies) to the AS or AW.

We emphasize that (9.6) contains *positive* wave numbers only, and spectral components move in the positive direction. Therefore, *the AW is the AS for a wave running in one direction.* So, positive definiteness of the spectrum (Property 3 AS, Section 2.2) results in waves running in one direction. In general, we introduce AWs in both directions:

$$w_\pm(x, t) = \frac{1}{\pi^2} \int_0^\infty \int_0^\infty U_\pm(k, \omega)e^{-i(\omega t \mp kx)}\, dk\, d\omega \tag{9.7}$$

and general complex waves are superpositions of w_+ and w_-. Containing the waves in both directions, they are the AS but not the AW.

Running Spectra For a given medium, k and ω are dependent variables connected with (9.3). Therefore, for the AW running in one direction, (9.6) takes the form

$$\begin{aligned} w(x, t) &= \frac{1}{\pi^2}\int_0^\infty\int_0^\infty U(k, \omega)\delta[k - k(x, \omega)]e^{-i(\omega t - kx)}\, dk\, d\omega \\ &= \frac{1}{\pi^2}\int_0^\infty U[k(x, \omega), \omega]e^{-i(\omega t - k(x,\omega)x)}\, d\omega \\ &= \frac{1}{\pi}\int_0^\infty W(x, \omega)e^{-i\omega t}\, d\omega \end{aligned} \tag{9.8}$$

where

$$W(x, \omega) = \frac{1}{\pi}U[k(x, \omega), \omega]e^{ik(x,\omega)x} \quad \text{for } \omega > 0$$

is the spectrum of the AW at a point x. Substituting (9.8) into (9.1) and using (9.3) again, we find an equivalent equation for the spectrum (see Problem 9.1 for more strictness):

$$\frac{\partial^2 W}{\partial x^2} + k^2(x, \omega)\, W = 0 \tag{9.9}$$

This is the ordinary differential equation because, for a fixed ω, $\partial/\partial x$ can be replaced by d/dx.

For a *uniform* medium with k independent of x, the general solution of (9.9):

$$W(x, \omega) = C_+(\omega)e^{ixk(\omega)} + C_-(\omega)e^{-ixk(\omega)} \tag{9.10}$$

contains two spectra running in opposite directions, and $w_\pm$ in (9.7) are their Fourier transforms. Arbitrary initial spectra $C_\pm(\omega)$ at $x = 0$ defines arbitrary running waves. Equation (9.9) is typical for the WKB approximation [7], and for nonuniform media, its asymptotic solution will be given in Section 9.3.1.

9.2 The Wave Center and Duration

To a certain degree, a running wave can be represented by its traveling center and duration at each point. We will show that the center and duration are the same for real waves and their AW, and that velocity of the center generalizes group velocity in a medium. We also discuss a paradoxical phenomenon in damping media.

9.2.1 Center and Duration of a Signal

First, we consider a real signal $u(t)$ with spectrum $U(\omega) = A(\omega)e^{i\psi(\omega)}$ and group delay $\tau(\omega) = d\psi/d\omega$. We introduce the first and second moments of the signal and its duration as follows:

$$\bar{t} = \frac{\int_{-\infty}^{\infty} tu^2(t)\,dt}{\int_{-\infty}^{\infty} u^2(t)\,dt} \qquad \bar{t^2} = \frac{\int_{-\infty}^{\infty} t^2u^2(t)\,dt}{\int_{-\infty}^{\infty} u^2(t)\,dt} \qquad T^2 = \overline{(t-\bar{t})^2} = \bar{t^2} - \bar{t}^2 \tag{9.11}$$

where an overbar denotes averaging in time. Clearly, the first moment $\bar{t}$ is the time position of a center, and T is the effective (quadratic) duration of a signal.

Applying Parseval's equality (2.29) to the Fourier pairs $u(t) \leftrightarrow U(\omega) = Ae^{i\psi}$ and $itu(t) \leftrightarrow U'(\omega) = [A' + i\tau A]e^{i\psi}$, we find

$$\begin{aligned}
\int_{-\infty}^{\infty} tu^2(t)\,dt &= \frac{-i}{2\pi}\int_{-\infty}^{\infty} U'(\omega)U^*(\omega)\,d\omega = \frac{-i}{2\pi}\int_{-\infty}^{\infty}(A' + i\tau A)A\,d\omega \\
&= \frac{-i}{2\pi}\left\{\left.\frac{A(\omega)^2}{2}\right|_{-\infty}^{\infty} + i\int_{-\infty}^{\infty}\tau(\omega)A(\omega)^2\,d\omega\right\} \\
&= \frac{1}{2\pi}\int_{-\infty}^{\infty}\tau(\omega)A(\omega)^2\,d\omega
\end{aligned}$$

$$\begin{aligned}
\int_{-\infty}^{\infty} t^2u^2(t)\,dt &= \frac{1}{2\pi}\int_{-\infty}^{\infty}|U'(\omega)|^2\,d\omega = \frac{1}{2\pi}\int_{-\infty}^{\infty}|A' + i\tau A|^2\,d\omega \\
&= \frac{1}{2\pi}\int_{-\infty}^{\infty}[A'(\omega)^2 + \tau(\omega)^2A(\omega)^2]\,d\omega
\end{aligned}$$

Therefore, relations (9.11) take the forms

$$\begin{aligned}
\bar{t} &= \frac{\int_{-\infty}^{\infty}\tau(\omega)A(\omega)^2\,d\omega}{\int_{-\infty}^{\infty}A(\omega)^2\,d\omega} = \bar{\bar{\tau}} \\
\bar{t^2} &= \frac{\int_{-\infty}^{\infty}\left(\frac{A'(\omega)^2}{A(\omega)^2} + \tau(\omega)^2\right)A(\omega)^2\,d\omega}{\int_{-\infty}^{\infty}A(\omega)^2\,d\omega} = \overline{\overline{\left(\frac{A'}{A}\right)^2}} + \overline{\overline{\tau^2}} \\
T^2 &= \overline{\overline{\left(\frac{A'}{A}\right)^2}} + \overline{\overline{\tau^2}} - \bar{\bar{\tau}}^2 = \overline{\overline{\left(\frac{A'}{A}\right)^2}} + \overline{\overline{(\tau - \bar{\bar{\tau}})^2}}
\end{aligned} \tag{9.12}$$

Here, a double overbar denotes averaging in frequency over the amplitude spectrum $A(\omega)^2$.

So, the time position of the center is the averaged group delay, and duration depends on amplitude and phase variations of the spectrum. Moreover, amplitude and phase components are summed in quadrature without interaction. Note also that (9.12) are the dual Fink's formulas (2.30).

Relation to the AS For real signals, $A(\omega)$ and $\tau(\omega)$ are even functions, and we can replace the limits in (9.12) by $0, \infty$. Then we come to the AS (2.12), and formulas (9.12) define its center and duration as well. So, besides (9.11), we also have

$$\bar{t} = \frac{\int_{-\infty}^{\infty} t|w(t)|^2\, dt}{\int_{-\infty}^{\infty} |w(t)|^2\, dt} \qquad \overline{t^2} = \frac{\int_{-\infty}^{\infty} t^2|w(t)|^2\, dt}{\int_{-\infty}^{\infty} |w(t)|^2\, dt} \tag{9.13}$$

and a real signal $u(t)$ and its AS $w(t)$ have the same center and duration. Since the AW is the AS for a real running wave, this is also true for the AW $w(x, t)$ at a fixed x.

We should make an additional comment [8]. According to (2.13), the spectrum $W(\omega)$ of the AS is discontinuous at $\omega = 0$ if $U(0) \neq 0$. Then its derivative $W'(\omega)$ contains the δ-function $U(0)\delta(\omega)$, and the integral

$$\int_{-\infty}^{\infty} t^2|w(t)|^2\, dt = \frac{1}{2\pi}\int_{-\infty}^{\infty} |W'(\omega)|^2\, d\omega$$

diverges even if the second moment in (9.11) converges. A constant displacement $U(0)$ is indistinctive for wave processes, however, and we set $U(0) = 0$. Then the second moment of the AS exists and, as has been shown, the center and duration of $w(t)$ are the same as those of $u(t)$.

9.2.2 Pure Dispersion

In uniform media, according to (9.10), the running spectrum is

$$W(x, \omega) = C(\omega)e^{ixk(\omega)}$$

and in general, the *complex* wave number $k(\omega) = k_r(\omega) + ik_i(\omega)$ defines dispersion and damping for each frequency. Then the amplitudes, phases, and group delays of the running spectrum are as follows:

$$A(x, \omega) = A_0(\omega)e^{-xk_i(\omega)} \qquad \psi(x, \omega) = \psi_0(\omega) + xk_r(\omega) \tag{9.14}$$

$$\tau(x, \omega) = \frac{\partial \psi}{\partial \omega} = \frac{d\psi_0}{d\omega} + x\frac{dk_r}{d\omega} = \tau_0(\omega) + x\tau_k(\omega) \tag{9.15}$$

Here, $A_0(\omega)$ and $\psi_0(\omega)$ are the amplitudes and phases of the initial spectrum $C(\omega)$. With distance, they are transformed into $A(x, \omega)$ and $\psi(x, \omega)$. Also, $\tau_0(\omega)$ is the group delay in the initial spectrum, whereas $\tau_k(\omega)$ is that in a medium per unit distance. (Dimensions of τ_0 and τ_k are sec and sec/m, respectively.)

For a pure dispersive medium of $k_i(\omega) = 0$, amplitudes are conserved, so that $A(x, \omega) = A_0(\omega)$. Then, from (9.15) and (9.12), we obtain the time position of the center at a point x:

$$\bar{t}(x) = \overline{\overline{\tau_0}} + x\overline{\overline{\tau_k}} \tag{9.16}$$

where $\overline{\overline{\tau_0}}$ and $\overline{\overline{\tau_k}}$ *averaged over the initial spectrum* $A_0(\omega)$ are independent of x. Therefore, whatever dispersion, the wave center moves *uniformly* with a constant averaged group delay per unit distance. It is well known that *narrowband* signals move by small distances with the group speed for a carrier frequency. Exact relation (9.16) generalizes this for any signals and distances.

Dispersion distorts the wave, however, and its duration is varying with distance. Using (9.15) and (9.12), we also obtain

$$T^2(x) = T_0^2 - \frac{\rho^2}{\Delta T_k^2} + \left(x\Delta T_k + \frac{\rho}{\Delta T_k}\right)^2 \tag{9.17}$$

where

$$T_0^2 = \overline{\overline{\left(\frac{A_0'}{A_0}\right)^2}} + \overline{\overline{(\tau_0 - \overline{\overline{\tau_0}})^2}} \qquad \Delta T_k^2 = \overline{\overline{(\tau_k - \overline{\overline{\tau_k}})^2}} \qquad \rho = \overline{\overline{\tau_0 \tau_k}} - \overline{\overline{\tau_0}} \cdot \overline{\overline{\tau_k}}$$

Here, T_0 is effective duration of the initial signal, ΔT_k is variation of duration due to dispersion (per unit distance), and ρ is the correlation factor between the initial group delay and that in the medium. Everywhere, averaging is done over the initial spectrum $A_0(\omega)$.

The correlation factor may be of either sign. If $\rho < 0$, dispersion compensates for phases of the initial spectrum. Then, the signal shortens, and its duration achieves a minimum at $x_0 = -\rho/\Delta T_k^2$. For $x > x_0$, phases are overcompensated, and duration grows again. The negative ρ corresponds to delay opposite to frequency modulation that is typical for time compression (Section 9.3.2).

Nonuniform Media A zeroth-order approximation (9.23) given below shows that, in dispersive *nonuniform* media, amplitudes are also conserved, $A(x, \omega) = A_0(\omega)$, whereas the phase and group delay are given by

$$\psi(x, \omega) = \psi_0(\omega) + \int_0^x k(x, \omega)\, dx \qquad \tau(x, \omega) = \tau_0(\omega) + \int_0^x \tau_k(x, \omega)\, dx$$

Then, averaging τ in (9.12), we obtain instead of (9.16):

$$\bar{t}(x) = \overline{\overline{\tau_0}} + \int_0^x \overline{\overline{\tau_k}}(x)\,dx \quad \text{whence} \quad \frac{d\bar{t}}{dx} = \overline{\overline{\tau_k}}(x) \quad \text{and} \quad v(x) = \frac{dx}{d\bar{t}} = \frac{1}{\overline{\overline{\tau_k}}(x)} \tag{9.18}$$

Thus, the local group delay (for a fixed x) *averaged in frequency* defines velocity of the center. In optics, this is the velocity of a light pulse depending on a refractive index at each point. In quantum mechanics, this is the velocity of a classical particle (see Section 9.4.2).

9.2.3 Damping and Causality

In damping media, spectral amplitudes $A(x, \omega)$ are varying. Then the averaged group delay depends on x, and the motion may be paradoxical.

Let the initial real signal be time-limited so that $u(0, t) = 0$ for $t < 0$. Then, because of relativistic causality, $u(x, t) = 0$ for $t < x/c$, where c is the light speed. Therefore, using the law of the mean, we find from the first equation (9.11):

$$\bar{t}(x) = \frac{\int_{x/c}^{\infty} tu^2(x, t)\,dt}{\int_{x/c}^{\infty} u^2(x, t)\,dt} \geq \frac{x}{c} \quad \text{so that} \quad \frac{x}{\bar{t}(x)} \leq c$$

Thus, causality limits the *mean velocity* $x/\bar{t}$ but not the *instantaneous velocity* $dx/d\bar{t}$ of a center. Under a causality condition, the instantaneous velocity may exceed c or be opposite to the mean velocity.

We now calculate the initial velocity of a center. Using $A(x, \omega)$ from (9.14), we differentiate (9.16) and (9.12) with respect to x and, for $x = 0$, obtain

$$\frac{1}{v(0)} = \frac{d\bar{t}(0)}{dx} = \overline{\overline{\tau_k}} - 2(\overline{\overline{\tau_0 k_i}} - \overline{\overline{\tau_0}} \cdot \overline{\overline{k_i}}) \tag{9.19}$$

The paradox is that the velocity $v(0)$ may be infinite or negative depending on the correlation factor between τ_0 and k_i. This behavior is easy to explain, however.

Let a wideband chirp signal be applied to a narrowband medium, and let only its beginning or end come through. Then, the signal shortens, and its center is shifted in time. Such a "motion" results not from propagation but from suppression of part of the signal. The end or beginning of the signal disappears, and therefore, the center of the transmitted signal is shifted in time backward or forward, respectively. If this shift compensates for the group delay in the medium, the resulting delay is zero, and the velocity $v(0)$ is infinite.

Pure Damping For pure damping, the wave number is imaginary, so that $k = ik_i(\omega)$, $k_r(\omega) = 0$, and $\tau_k(\omega) = 0$. Then, spectral components are weakened but not delayed, and their phases are changeless in the medium. Because of selective damping, the signal is distorted. Its maximum, however (when the components are in phase), remains at the same time point for any x. Therefore, "propagation" takes no time, and possibly, we should not interpret such processes as waves.

In general, the wave number may be imaginary for part of a spectrum and real for another part. Then, a signal is partly weakened and partly delayed. This leads to considerable distortions that can be confused with a loss of causality. The paradox of tunneling is of this kind (Section 9.4.3).

9.3 The Wave Shape

We have considered the center and duration of a running wave but not its shape. Now we develop two approximate methods for the shape. First an asymptotic solution will be constructed for the spectrum running in a slowly varying medium. Next, we generalize Whitham's method for defining not only frequency but also amplitude of a wave.

9.3.1 Asymptotic Solution

Equation (9.9) is formally the same as for the adiabatic pendulum. As in Section 6.1.1, for real k^2 (this case includes dispersion and damping because k may be real or imaginary), we multiply (9.9) by W^* and take the imaginary part. Writing the spectrum as $W(x, \omega) = A(x, \omega)e^{i\psi(x,\omega)}$, we denote $\chi = \partial\psi/\partial x$ and obtain

$$\mathrm{Im}\left(\frac{\partial^2 W}{\partial x^2} W^* + k^2|W|^2\right) = \mathrm{Im}\left(\frac{\partial^2 W}{\partial x^2} W^*\right) = \frac{\partial}{\partial x}\,\mathrm{Im}\left(\frac{\partial W}{\partial x} W^*\right)$$

$$= \frac{\partial}{\partial x}\,\mathrm{Im}\left(\frac{\partial A}{\partial x} A + i\chi A^2\right) = \frac{\partial}{\partial x}(\chi A^2) = 0$$

So, $\chi A^2 = f(\omega)$ is *independent of* x, and the running spectrum that satisfied (9.9) takes the form

$$W(x, \omega) = Ae^{i\psi} = \sqrt{\frac{f(\omega)}{\chi(x, \omega)}}e^{i\psi} = C(\omega)\sqrt{\frac{\chi(0, \omega)}{\chi(x, \omega)}}e^{i\int_0^x \chi(x,\omega)\,dx} \quad (9.20)$$

where $f(\omega)$ is found from the initial spectrum $C(\omega)$ at $x = 0$.

The function χ is still unknown, but substituting (9.20) into (9.9), we come to the equation

$$\chi^2 + \frac{1}{2}\frac{\chi''}{\chi} - \frac{3}{4}\left[\frac{\chi'}{\chi}\right]^2 = k^2 \tag{9.21}$$

where prime denotes derivation with respect to x. Now we suppose that the wave number is slowly varying, $k = k(\varepsilon x, \omega)$ with $\varepsilon \ll 1$. Then, χ also depends on εx, and the derivatives in (9.21) are of the order ε^2. Neglecting them in the first order, we have $\chi = \pm k$, and the second-order solution results by iteration. Finally, we obtain

$$\chi = \begin{cases} \pm k & \text{for the first order} \\ \pm\left[k - \dfrac{1}{4}\dfrac{k''}{k} + \dfrac{3}{8}\left(\dfrac{k'}{k}\right)^2\right] & \text{for the second order} \end{cases} \tag{9.22}$$

Clearly, the medium must be slowly varying with respect to a wavelength, and continuing the iterations, we can find higher corrections to χ. In the first order, for $\chi = \pm k$, (9.20) is the WKB approximation, but this relation is exact for the function χ satisfied (9.21).

In contrast with (9.10), not only phases but also amplitudes of the spectrum are varying in nonuniform dispersive media. (9.20) and (9.22) define the spectrum, and the AW $w(x, t)$ is obtainable with the Fourier transform (9.8). So, the FFT becomes an effective numerical method for the waves.

Zeroth-Order Solution For $\chi = k$, from (9.20) we find

$$\frac{\partial W}{\partial x} = \left(-\frac{k'}{2k} + ik\right) W$$

and the zeroth-order equation results if we neglect $k' \sim \varepsilon$ compared with k^2. Then we obtain

$$\frac{\partial W}{\partial x} = ik(x, \omega)\, W \quad \text{and} \quad W(x, \omega) = C(\omega) e^{i \int_0^x k(x,\omega)\, dx} \tag{9.23}$$

So, in zeroth order, spectral amplitudes are conserved even for nonuniform dispersive media. Useful relations can be found within a framework of this approximation, and its accuracy is often acceptable (Section 9.4.3).

9.3.2 Whitham's Method

This method gives another opportunity, and its basic idea is easy [2]. A real harmonic wave is $u(x, t) = \cos(\omega t - kx)$, where ω and k are constants connected with the dispersion relation

$$k = k(\omega) \tag{9.24}$$

Whitham writes a general real wave in the form (9.5) and defines its *local frequency* and *local wave number* as $\omega(x, t) = \partial\phi/\partial t$, $k(x, t) = -\partial\phi/\partial x$. His crucial assumption is that the *slowly varying* local k and ω are connected with the same relation (9.24). That leads to equations for $\omega(x, t)$ or $k(x, t)$. This approach relates to global and local frequency notions and may be reasserted with the stationary phase approximation (Problem 9.6).

So, Whitham's method is an approximate quasistationary approach replacing global spectral frequency ω by the local instantaneous frequency $\omega(x, t)$ and global wave number k by local $k(x, t)$. Therefore, this method implicitly uses the AW. As we have seen in Section 3.5, only the AS provides slow frequencies for real signals, and the AW does that for ω and k. Using the AW properly, we now generalize the method for amplitudes.

First, writing the AW in the form

$$w(x, t) = a(x, t)e^{i\phi(x,t)} = e^{i\theta(x,t)}$$

we introduce its complex phase and complex local ω and k as in Section 3.6:

$$\theta(x, t) = \phi(x, t) - i\ln[a(x, t)] \qquad \omega(x, t) = \frac{\partial\theta}{\partial t} \qquad k(x, t) = -\frac{\partial\theta}{\partial x} \tag{9.25}$$

Further, following Whitham, we assume that the slowly varying complex k and ω are connected with (9.24). Then, from (9.25), we have

$$\frac{\partial^2\theta}{\partial x\partial t} = \frac{\partial\omega}{\partial x} = -\frac{\partial k}{\partial t} \quad \text{so that} \quad \frac{\partial\omega}{\partial x} + \frac{\partial k}{\partial t} = 0$$

where in view of (9.24), $\partial k/\partial t = k'(\omega)\partial\omega/\partial t$. Therefore, we come to Whitham's equation

$$\frac{\partial\omega}{\partial x} + \tau(\omega)\frac{\partial\omega}{\partial t} = 0 \tag{9.26}$$

where $\tau(\omega) = k'(\omega)$ is the group delay (9.4). Originally, this nonlinear equation has been derived for a real frequency, but now it defines the amplitude of the AW also (more accurately, its logarithmic derivative).

Characteristics Now we point out a method for solving (9.26). Characteristics $x(t)$ are the curves on the (x, t)-plane for which the function $\omega(x, t)$ given by the equation is constant, and they are defined by

$$\frac{dx}{dt} = \frac{1}{\tau(\omega)} \tag{9.27}$$

Really, for the characteristic $x(t)$, from (9.27) and (9.26) we have

$$\frac{d\omega}{dt} = \frac{\partial\omega}{\partial x}\frac{dx}{dt} + \frac{\partial\omega}{\partial t} = \frac{1}{\tau(\omega)}\left[\frac{\partial\omega}{\partial x} + \tau(\omega)\frac{\partial\omega}{\partial t}\right] = 0$$

So, $\omega(x, t)$ is constant. Therefore, $\tau(\omega)$ and dx/dt are also constant, and the characteristic is the straight line given by

$$t = \xi + x\tau[\omega(\xi)] \tag{9.28}$$

Here, ξ is the time point where the characteristic intersects the axis $x = 0$. Since ω is constant in the characteristic, we take it at ξ and write it as $\omega(\xi)$.

We have solved (9.26). At $x = 0$, the initial complex frequency $\omega(\xi)$ is given, and it includes the initial amplitude, too. The dispersion relation $\tau(\omega)$ is also given, and (9.28) implicitly determines the dependence $\xi = \xi(x, t)$. Substituting this dependence into $\omega(\xi)$, we find the frequency $\omega(x, t)$ at any x and t (see Problem 9.2 for a formal proof). Finally, for the complex $\omega(x, t)$ obtained, we find the complex phase and the AW as follows:

$$\theta(x, t) = \int \omega(x, t)\, dt + g(x) \qquad w(x, t) = e^{i\theta(x, t)} \tag{9.29}$$

and for pure dispersion, the unknown function $g(x)$ should be found from energy conservation. Then $|w(x, t)|$ defines the amplitude of a running wave while its frequency is a real part of the complex $\omega(x, t)$.

Time Compression As illustrated in Figure 9.1, each narrow strip of a chirp signal is delayed by $x\tau(\omega)$ according to its frequency $\omega(t)$. Therefore, for the linear group delay opposite to frequency modulation, all strips will arrive at some point x at one instant. Then the duration shortens, the amplitude grows, and the signal gets compressed in time at that point.

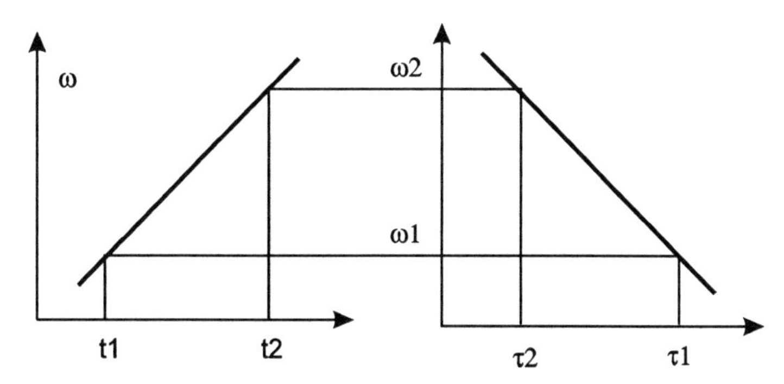

Figure 9.1 Compression of a chirp signal.

This clarifies the idea but not the amplitude and duration of a wave. Whitham's method does that easily, however. For a chirp signal with $\omega(\xi) = \omega_0 + \beta\xi$ and with Gaussian amplitude $a(\xi) = e^{-\xi^2/2}$, the complex frequency is $\omega(\xi) = \omega_0 + (\beta + i)\,\xi$. Then, for a linear dispersion $\tau(\omega) = \tau_1 - (\omega - \omega_0)$, equation (9.28) takes the form

$$t = \xi + x[\tau_1 - (\beta + i)\xi]$$

and defines

$$\xi(x, t) = \frac{t - x\tau_1}{1 - x(\beta + i)}$$

Therefore, the complex frequency and phase are as follows:

$$\omega(x, t) = \omega_0 + (\beta + i)\frac{t - x\tau_1}{1 - x(\beta + i)}$$

$$\theta(x, t) = \omega_0 t + (\beta + i)\frac{(t - x\tau_1)^2}{2[1 - x(\beta + i)]}$$

Finally, normalizing the AW $w = e^{i\theta}$ to conserve energy at each x, we find the amplitude and duration as follows:

$$a(x, t) = \frac{1}{T(x)} \exp\left[-\frac{(t - x\tau_1)^2}{2T^2(x)}\right]$$

$$T^2(x) = \frac{1}{\beta^2 + 1} + (\beta^2 + 1)\left[\frac{\beta}{\beta^2 + 1} - x\right]^2$$

The amplitude is shown in Figure 9.2, and the signal is compressed at $x = \beta/(\beta^2 + 1)$ where duration has a minimum. The same duration results from (9.17).

Whitham's Method for Nonuniform Media Since derivatives of k and ω are neglected, Whitham's method is of zeroth order like (9.23). For the group delay depending on x, (9.27) takes the form $dx/dt = 1/\tau(x, \omega)$, and characteristics are not straight lines but obey the equation

$$t = \xi + \int_0^x \tau[x, \omega(\xi)]\, dx \tag{9.30}$$

As before, ξ is the point where the characteristic intersects the axis $x = 0$. Solving (9.30) for $\xi(x, t)$, one can find the frequency and amplitude of the AW in a nonuniform medium (see Section 9.4.1). The FFT of the spectrum (9.20) determines the AW more accurately, however. A numerical solution of (9.30) is not easier than the FFT, and practically, Whitham's method is mostly expedient for analytical solutions in simple cases.

9.4 Quantum Mechanical Wave Packets

Now we address Schrödinger's equation

$$\frac{\hbar^2}{2m}\frac{\partial^2 \psi}{\partial x^2} + i\hbar\frac{\partial \psi}{\partial t} - V(x)\psi = 0 \tag{9.31}$$

which defines the quantum mechanical wave function $\psi(x, t)$ for a particle of mass m in a field of potential $V(x)$. We will show that the wave center, as defined

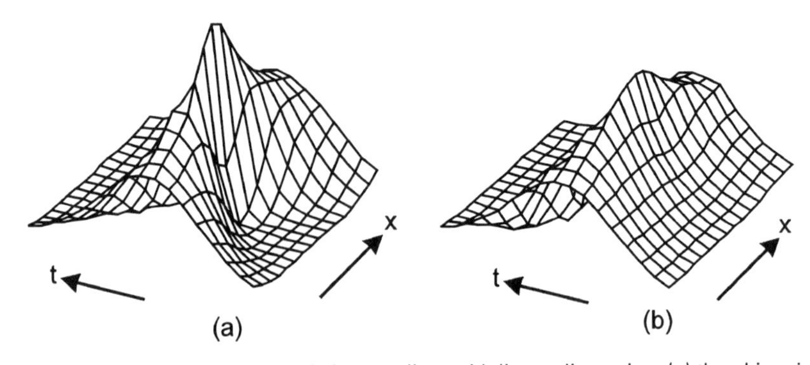

Figure 9.2 Amplitude of chirp signals in a medium with linear dispersion: (a) the chirp signal with the complex frequency $\omega(t) = (5 + i)t$ and (b) the chirp signal with the distorted frequency $\omega(t) = (5 + i)t - 2t^2$.

in Section 9.2, represents the associated classical particle. Besides, *complex* wave functions $\psi(x, t)$ are often AS and, for particles moving in one direction, they are also AW. That results not from our definition of amplitude and frequency but from the equation itself. Therefore, possibly, quantum mechanics provides a physical basis of the AS and AW.

For (9.31), the general dispersion relation (9.3) takes the form

$$k(x, \omega) = \pm\frac{\sqrt{2m}}{\hbar}\sqrt{\hbar\omega - V(x)}, \qquad \tau(x, \omega) = \frac{\partial k}{\partial \omega} = \pm\sqrt{\frac{m}{2[\hbar\omega - V(x)]}} \tag{9.32}$$

and typical potential functions are shown in the Figure 9.3.

Because of the dependence on x of the potential $V(x)$, the wave packet ψ is moving in a nonuniform medium. If $\hbar\omega > V(x)$, k and τ are real, and ψ is propagating in a dispersive medium with the group delay decreasing at high frequencies according to (9.32). If $\hbar\omega < V(x)$, then k and τ are imaginary. Then, dispersion is replaced by damping, and a wave packet is tunneling through a potential barrier. In both cases, k^2 is real, and the method of Section 9.3.1 is applicable. Note also that a wideband wave packet may be partly propagating and partly damping. For a free particle, when $V(x) = 0$, the medium is uniform, and k is independent of x.

9.4.1 Center of a Wave Packet and a Classical Particle

For $\hbar\omega > V(x)$, the center of a wave packet represents a classical particle. In view of (9.12) and according to (9.18) and (9.32), velocity of the center at a point x is given by

$$v(x) = \frac{1}{\overline{\overline{\tau}}(x)} = \frac{\int A^2(\omega)d\omega}{\int\sqrt{\frac{m}{2(\hbar\omega - V(x))}}A^2(\omega)\,d\omega} \approx \sqrt{\frac{2(\hbar\omega_0 - V(x))}{m}} \tag{9.33}$$

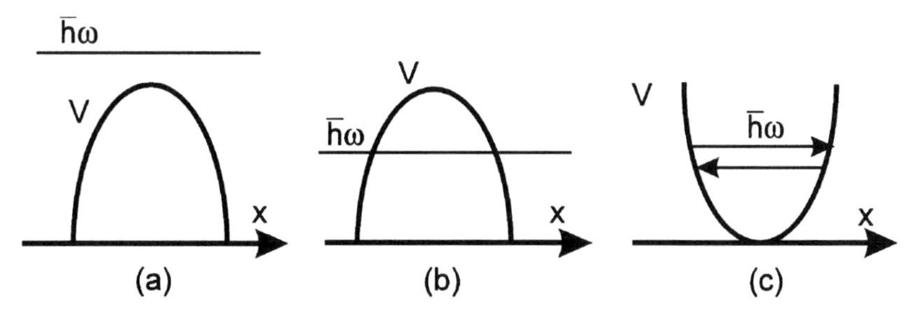

Figure 9.3 Potential functions: (a) dispersive motion, (b) tunneling, and (c) oscillations.

and therefore

$$\hbar\omega_0 = \frac{mv^2}{2} + V(x)$$

Here, for a *narrowband* spectrum $A(\omega)$ concentrated around ω_0, we have replaced ω by ω_0. So, averaging frequency (energy) components, we obtain the classical velocity of a particle as the velocity of a wave center and the classical energy balance for the total energy $E = \hbar\omega_0$.

We now apply Whitham's method to (9.31). In view of (9.32), the equation of characteristics (9.30) takes the form

$$t = \xi + \sqrt{\frac{m}{2}} \int_0^x \frac{dx}{\sqrt{\hbar\omega(\xi) - V(x)}} \tag{9.34}$$

This equation has been derived for a wave, but it is the integral of (9.33) and describes a classical motion. In fact, a classical particle starting from $x = 0$ at the instant ξ with the energy $\hbar\omega(\xi)$ achieves the point x at the instant t given by (9.34). Nevertheless, this equation also defines the wave function.

If we specify potential in the Figure 9.3(a) as

$$V(x) = V_0\left(1 - \frac{x^2}{x_0^2}\right) \quad \text{for } -x_0 < x < x_0 \tag{9.35}$$

we find that (9.34) takes the form

$$\begin{aligned} t &= \xi + \sqrt{\frac{mx_0^2}{2V_0}}\left[\operatorname{arsh}\left(\sqrt{\frac{V_0}{\hbar\omega(\xi) - V_0}}\right) + \operatorname{arsh}\left(\frac{x}{x_0}\sqrt{\frac{V_0}{\hbar\omega(\xi) - V_0}}\right)\right] \\ &\approx \xi + (x + x_0)\sqrt{\frac{m}{2[\hbar\omega(\xi) - V_0]}} \end{aligned} \tag{9.36}$$

For a Gaussian initial wave packet:

$$\psi_0(\xi) = e^{i\omega_0\xi - \xi^2/2T_0^2} \quad \text{at } x = -x_0$$

the complex frequency is $\omega(\xi) = \omega_0 + i\xi/T_0^2$, and for $\xi/T_0^2 \ll \omega_0 - V_0/\hbar$, equation (9.36) can be solved in the same way as in Section 9.3.2 (see also Problem 9.8). Then, we find that duration of the wave packet is increasing with distance:

$$T^2(x) \approx T_0^2 + \frac{(x + x_0)^2\hbar^2 m}{8T_0^2(\hbar\omega_0 - V_0)^3} \tag{9.37}$$

For $V_0 = 0$, duration for a free particle is also increasing (see Problems 9.3 and 9.4).

9.4.2 Paradox of Tunneling

Why does duration widen in (9.37)? Components of various frequencies are delayed by various times. Therefore, though the initial packet is modulated in amplitude only, frequency modulation appears that widens the spectrum. The spectrum is to be conserved (for pure dispersion considered), however, and duration increases for its narrowing. On the other hand, from a corpuscular viewpoint, particles of various energy (frequency) moving with various speeds disperse over a wide range in time (for a fixed x) or space (for a fixed t).

When a free particle moves toward a barrier, the wave packet broadens, and since the group delay is less at high frequencies, frequency modulation appears with high frequencies at the beginning and low frequencies at the end of the packet. When tunneling, $\hbar\omega < V(x)$, and the wave number is imaginary for a low-frequency part of the spectrum. This part is stopped whereas a high-frequency part passes above (or through) the barrier.

Assuming the initial chirp packet with Gaussian amplitude and the potential (9.35), we computed the spectra (9.20) and the wave packets inside the barrier (with the FFT). Because of suppression of low frequencies, the initial spectrum is narrowed (Figure 9.4(a)). The high frequencies above the barrier reside at the early part of the packet, and therefore, the packet shortens while its center is shifted in time toward the beginning (Figure 9.4(b)). Using (9.13), we also computed time positions of the center for the wave packets found in first and zeroth orders (with (9.23) for zeroth order).

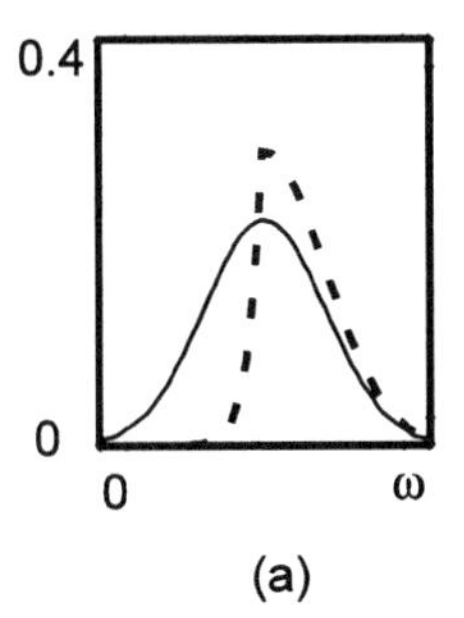

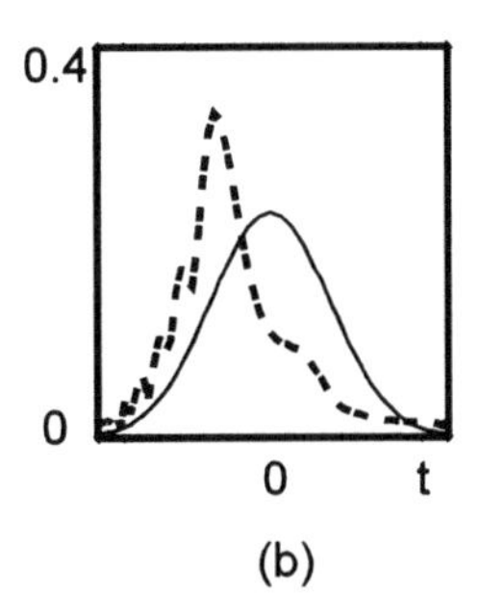

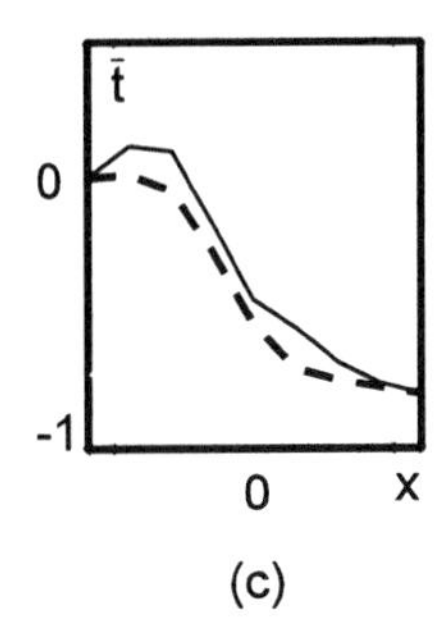

Figure 9.4 Tunneling of a chirp wave packet through a barrier: (a), (b) are initial (solid lines) and final (dotted lines) spectra and amplitudes; (c) is the time position of the wave center for the first (solid line) and zeroth (dotted line) orders. The zeroth-order solution is of acceptable accuracy. Spectra and wave packets are normalized, and overall damping is ignored.

As shown in Figure 9.4(c), the center of the transmitted packet leaves the barrier before the center of the incident packet has arrived. That is often understood as a loss of causality. This paradox has given birth to many alternative approaches, and time of tunneling through a barrier is still an open question [4].

We believe that tunneling itself takes no time (see Section 9.2.3), and from a corpuscular viewpoint, we may argue in another way. Particles are moving with various speeds, and only the faster ones of higher energy have a chance of overcoming the barrier. Therefore, the mean time of arrival of transmitted particles is less than that of all incident particles. This mean time is presented by the wave center.

Thus, the paradox arises if we associate the wave packet with a single particle. The wave packet generally represents an ensemble of particles, however, and the barrier is merely a filter for fast (high-frequency) particles arriving before the others. From a wave viewpoint, this is just the same paradox as in Section 9.2.3, and no physical conflict emerges.

9.4.3 Relation to the AS and AW

We now show that wave packets are often the AS and AW. A rigorous method for solving Schrödinger's equation is separation of variables when we seek $\psi(x, t)$ in the form

$$\psi(x, t) = \sum_n c_n T_n(t) X_n(x)$$

where orthogonal (for unequal n) functions $T_n(t)$ and $X_n(x)$ depend on t and x separately. Due to linearity, each product $T_n(t)X_n(x)$ obeys the equation, and substituting into (9.31), we find

$$\frac{1}{X_n}\left(-\frac{\hbar^2}{2m}\frac{d^2 X_n}{dx^2} + V(x)X_n\right) = \frac{i\hbar}{T_n}\frac{dT_n}{dt}$$

Here, the left-hand side depends on x, whereas the right-hand side depends on t. Therefore, both equal to a constant λ_n, and we obtain two equations:

$$-\frac{\hbar^2}{2m}\frac{d^2 X_n}{dx^2} + V(x)X_n = \lambda_n X_n \qquad \frac{dT_n}{dt} = -i\frac{\lambda_n}{\hbar}T_n \tag{9.38}$$

The first equation (9.38) shows that $X_n(x)$ and λ_n are the eigenfunctions and eigenvalues of the Hamiltonian operator $\mathcal{H} = -\frac{\hbar^2}{2m}\frac{d^2}{dx^2} + V(x)$. The second equation defines the $T_n = e^{-i\lambda_n t/\hbar}$, and finally we have

$$\psi(x, t) = \sum_n c_n X_n(x)\, e^{-i\omega_n t} \qquad \omega_n = \frac{\lambda_n}{\hbar} \tag{9.39}$$

So, the frequency spectrum of a wave function is defined by the eigenvalues λ_n. Multiplying the first equation (9.38) by X_n^* and integrating by parts, we also find

$$\begin{aligned}\lambda_n \int |X_n|^2\, dx &= \int \left[-\frac{\hbar^2}{2m} \frac{d^2 X_n}{dx^2} + V(x) X_n \right] X_n^*\, dx \\ &= \int \left[\frac{\hbar^2}{2m} \left| \frac{dX_n}{dx} \right|^2 + V(x) |X_n|^2 \right] dx\end{aligned}$$

This shows that, for positive potentials $V(x) \geq 0$, the eigenvalues λ_n are *positive.* Then the frequencies ω_n are also positive, and the wave function (9.39) is the AS.

The condition $V(x) \geq 0$ is not met for the Coulomb potential $V(x) = -1/|x|$ and some others, so that wave functions are not ASs generally. However, the condition is met for quantum oscillators and other potentials in Figure 9.3. In the classical limit, this explains the AS for common oscillators. Under the same condition, the wave packets moving in one direction are also AW. An oscillating particle is coupled with a standing wave, however, which includes two AWs moving in opposite directions.

9.5 Nonlinear Waves

High-power signals vary physical properties of a medium, and nonlinear effects appear. Considering voltage-dependent permitivity, we now investigate (9.1) with a small cubic correction as follows:

$$\frac{\partial^2 u}{\partial x^2} - \frac{1}{c^2} L\left(\frac{\partial}{\partial t}\right) u = \frac{\varepsilon}{3c^2} L\left(\frac{\partial}{\partial t}\right) u^3 \tag{9.40}$$

Here, velocity of propagation c is explicitly pointed out. Studying this equation, we employ the AW and the frequency-separation procedure developed in Chapter 7 for nonlinear oscillations. We will show that nonlinearity results in velocity modulation and frequency modulation of propagating waves. If frequency modulation is compensated with dispersion in a medium, a solitary wave is propagating without distortions.

9.5.1 Frequency Separation

For $\varepsilon = 0$, the equation is linear, and we assume that a narrowband wave $u(x, t)$ is concentrated around the frequency $\omega_0 = 1$. Then, using notations from Section 7.1, we write $u^{(0)} = u_1$, and the first-order iterative equation (7.7) takes the form

$$\frac{\partial^2 u^{(1)}}{\partial x^2} - \frac{1}{c^2} L\left(\frac{\partial}{\partial t}\right) u^{(1)} = \frac{\varepsilon}{3c^2} L\left(\frac{\partial}{\partial t}\right) u_1^3 \tag{9.41}$$

Now, again as in Section 7.1, we transfer to the AW by setting $u = (w + w^*)/2$ in both sides and separating the terms at positive frequencies. Then, we have

$$\frac{\partial^2 w^{(1)}}{\partial x^2} - \frac{1}{c^2} L\left(\frac{\partial}{\partial t}\right) w^{(1)} = \frac{\varepsilon}{12c^2} L\left(\frac{\partial}{\partial t}\right) (w_1^3 + 3a^2 w_1) \qquad a^2 = w_1 w_1^* \tag{9.42}$$

The two addends in the right-hand side are at the frequencies $\omega = 1$ and $\omega = 3$, and the same frequencies must be in the left-hand side. Therefore, we set $w^{(1)} = w_1 + w_3$, and after frequency separation, the first-order system takes the form

$$\frac{\partial^2 w_1}{\partial x^2} - \frac{1}{c^2} L\left(\frac{\partial}{\partial t}\right) \left[1 + \frac{\varepsilon}{4} a^2\right] w_1 = 0 \tag{9.43}$$

$$\frac{\partial^2 w_3}{\partial x^2} - \frac{1}{c^2} L\left(\frac{\partial}{\partial t}\right) w_3 = \frac{\varepsilon}{12c^2} L\left(\frac{\partial}{\partial t}\right) w_1^3 \tag{9.44}$$

So, the third harmonic appears in the first order, and higher harmonics will appear in higher orders. When propagating in a nonlinear medium, especially with equal velocities, interaction of harmonics may result in shock waves [2]. We assume, however, that in a *narrowband* medium, the first harmonic is only allowed, which is typical for fiberoptics. Then, (9.44) may be ignored, and (9.43) contains the correction appearing at each point from the nonlinear self-action of the first harmonic.

9.5.2 Velocity Modulation and Frequency Modulation

For slow amplitude, we neglect its derivatives and take εa^2 out of the operator L (with a total error of ε^2). Then, (9.43) takes the form

$$\frac{\partial^2 w_1}{\partial x^2} - \frac{1}{c^2(a)} L\left(\frac{\partial}{\partial t}\right) w_1 = 0 \quad \text{where } c(a) \approx c\left(1 - \frac{\varepsilon}{8} a^2\right) \tag{9.45}$$

shows that *velocity modulation* appears in a medium. We have seen that similar nonlinear interaction varies frequency of a pendulum or a generator that can be interpreted as a variation of the capacity of its resonant circuit. In a medium, variation of the capacity (permitivity) produces velocity modulation.

In a linear medium, the wave spectrum conserves or narrows but does not broaden. Velocity modulation is a nonlinear effect that widens the spectrum. For simplicity, we consider the medium without dispersion so that $L = \partial^2/\partial t^2$. Then, for a wave running in the positive direction, (9.45) can be reduced to the first-order equation as follows (see Problem 9.7):

$$\frac{\partial w}{\partial x} + \frac{1}{c(a)}\frac{\partial w}{\partial t} = 0 \tag{9.46}$$

This is just Whitham's equation (9.26) (for w instead of ω), and (9.28) for characteristics takes the form

$$t = \xi + \frac{x}{c}\left(1 + \frac{\varepsilon}{8}a^2(\xi)\right) \tag{9.47}$$

Its zeroth-order solution (for $\varepsilon = 0$) is $\xi = t - x/c$, and by iteration, the first-order solution is as follows:

$$\xi = t - \frac{x}{c} - \frac{\varepsilon x}{8c}a^2\left(t - \frac{x}{c}\right)$$

So, if the initial AM signal (at $x = 0$) is $w_0(\xi) = a(\xi)e^{i\omega_0\xi}$, the propagating signal takes the form

$$w(x, t) = a(\xi)\exp i\omega_0\left(\xi + \frac{\varepsilon x}{8c}a^2(\xi)\right) \quad \text{where } \xi = t - \frac{x}{c} \tag{9.48}$$

Around a top, the amplitude is $a^2(\xi) \approx a_0^2(1 - \xi^2)$, and linear frequency modulation appears as follows:

$$\omega(\xi) = \omega_0 - \frac{a_0^2\omega_0\varepsilon x}{4c}\xi$$

Thus, velocity modulation results in frequency modulation, and the longer distance x, the greater the frequency deviation and the wider spectrum of a wave.

9.5.3 Solitary Waves

Frequency modulation produces phase shifts. Dispersion in a medium also produces phase shifts. If they compensate each other, a *solitary wave* propagates without distortions in a nonlinear dispersive medium [5].

The phase compensation carries out at each point x, and we consider it for a small distance δx around $x = 0$. In the first order with respect to δx, the frequency modulated signal (9.48) can be written as

$$w(t) = [a(t) + i\mu\delta x a^3(t)]e^{i\omega_0 t} \qquad \mu = \frac{\omega_0 \varepsilon}{8c}$$

Its spectrum takes the form

$$\begin{aligned} W(\omega) &= A(\omega - \omega_0) + i\mu\delta x A_3(\omega - \omega_0) \\ &= A(\omega - \omega_0)\left[1 + i\mu\delta x \frac{A_3(\omega - \omega_0)}{A(\omega - \omega_0)}\right] \\ &= A(\omega - \omega_0)\left[1 + i\mu\delta x \frac{a_0^2}{\sqrt{3}} e^{(\omega - \omega_0)^2 T^2/3}\right] \end{aligned} \tag{9.49}$$

Here, $A(\omega)$ is the spectrum of $a(t)$, $A_3(\omega)$ is the spectrum of $a^3(t)$, and we have assumed the Gaussian amplitude $a(t) = a_0 e^{-t^2/2T^2}$. Then,

$$A(\omega) = \sqrt{2\pi}\, a_0 T e^{-\omega^2 T^2/2} \qquad A_3(\omega) = \sqrt{\frac{2\pi}{3}}\, a_0^3 T e^{-\omega^2 T^2/6}$$

that results in (9.49).

For a small δx, dispersion produces the phase shift as follows:

$$\begin{aligned} w(\delta x, t) &= \frac{1}{\pi}\int_0^\infty W_0(\omega) e^{i[\omega t - \delta x k_1(\omega)]}\, d\omega \\ &= \frac{1}{\pi}\int_0^\infty W_0(\omega)[1 - i\delta x k_1(\omega)] e^{i\omega t}\, d\omega \end{aligned}$$

where $k_1(\omega) = k(\omega) - \omega/c$ is the "dispersion part" of a wave number. Using (9.49) as the initial spectrum W_0, we find the condition of phase compensation as follows:

$$k_1(\omega) = \frac{\mu a_0^2}{\sqrt{3}} e^{(\omega - \omega_0)^2 T^2/3}$$

Therefore, around ω_0, the group delay must be linear:

$$\tau(\omega) = \frac{1}{c}\left[1 + \frac{\varepsilon\omega_0 a_0^2 T^2}{12\sqrt{3}}(\omega - \omega_0)\right] \tag{9.50}$$

A linear group delay is typical, and solitary waves propagate without distortions in nonlinear dispersive media. We have explained them using the AW and frequency separation. This amazing wave effect comes from interaction between dispersion and nonlinear frequency modulation.

9.6 Supplementary Problems

Problem 9.1

In Section 9.1, (9.9) for the running spectrum was not found rigorously. The point is that (9.1) is for a real wave $u(x, t)$ but not the AW $w(x, t)$. Therefore, we should *not* substitute (9.8) into (9.1). Show, however, that (9.9) is true.

Solution Let us repeat the development for a real wave $u(x, t)$. Because the wave number is frequency-dependent according to (9.3), for a real wave propagating in the positive direction, we have

$$\begin{aligned} u(x, t) &= \frac{1}{4\pi^2}\int_{-\infty}^{\infty}\int_{-\infty}^{\infty} U(k, \omega)e^{-i(\omega t - kx)}dk\, d\omega \\ &= \frac{1}{4\pi^2}\int_{-\infty}^{\infty}\int_{-\infty}^{\infty} U(k, \omega)\delta(k - k(x, \omega))e^{-i(\omega t - kx)}dk\, d\omega \\ &= \frac{1}{2\pi}\int_{-\infty}^{\infty} F(x, \omega)e^{-i\omega t}d\omega \end{aligned}$$

where

$$F(x, \omega) = \frac{1}{2\pi}U[k(x, \omega), \omega]e^{ik(x,\omega)x}$$

is the spectrum of $u(x, t)$ taken at a fixed x. Substituting this $u(x, t)$ into (9.1) is rigorous, and we come to (9.9) for $F(x, \omega)$. For $\omega > 0$, however, the spectrum $F(x, \omega)$ differs from the $W(x, \omega)$ in (9.8) by a constant factor only. Therefore, the same equation (9.9) is true for the running spectrum $W(x, \omega)$.

Problem 9.2

With a direct substitution, reassert that Whitham's equation (9.26) is satisfied with arbitrary function $\omega(\xi)$ if $\xi(x, t)$ meets equation of characteristics (9.28).

Solution For $\omega[\xi(x, t)]$, we have

$$\frac{\partial \omega}{\partial x} = \omega'(\xi)\frac{\partial \xi}{\partial x} \qquad \frac{\partial \omega}{\partial t} = \omega'(\xi)\frac{\partial \xi}{\partial t} \tag{9.51}$$

On the other hand, differentiating (9.28) with respect to x and t, we find

$$0 = \frac{\partial \xi}{\partial x} + \tau(\omega) + x\tau'(\omega)\omega'(\xi)\frac{\partial \xi}{\partial x} \qquad 1 = \frac{\partial \xi}{\partial t} + x\tau'(\omega)\omega'(\xi)\frac{\partial \xi}{\partial t}$$

Therefore

$$\frac{\partial \xi}{\partial x} = -\frac{\tau(\omega)}{1 + x\tau'(\omega)\omega'(\xi)} \qquad \frac{\partial \xi}{\partial t} = \frac{1}{1 + x\tau'(\omega)\omega'(\xi)}$$

Substituting these derivatives into (9.51) and then into (9.26), we come to the identity

$$\frac{\partial \omega}{\partial x} + \tau(\omega)\frac{\partial \omega}{\partial t} = \frac{\omega'(\xi)}{1 + x\tau'(\omega)\omega'(\xi)}[-\tau(\omega) + \tau(\omega)] = 0$$

which shows that the solution obtained with characteristics satisfies (9.26).

Problem 9.3

We have considered waves in a *time-frequency domain* taking t and ω as basic variables. Harmonic waves $e^{i(\omega t - kx)}$ are symmetrical with respect to t and x, however, and for a uniform medium, solution of (9.1) can be written with either Fourier integral:

$$\begin{aligned} u(x, t) &= \frac{1}{2\pi}\int_{-\infty}^{\infty} C(\omega)e^{i[xk(\omega)-\omega t]}\,d\omega \\ u(x, t) &= \frac{1}{2\pi}\int_{-\infty}^{\infty} U_0(k)e^{i[\omega(k)t-kx]}\,dk \end{aligned} \tag{9.52}$$

The first integral has been used in the text. In the second integral, $U_0(k)$ is the initial *spatial spectrum* given at $t = 0$, and the basic variables in a *space-wave number domain* are x and k. Then, dispersion relations

$$\omega = \omega(k) \qquad v(k) = \frac{d\omega}{dk} \tag{9.53}$$

define the frequency and group velocity as functions of k.

Point out equations for the wave center and for Whitham's method in the space-wave number domain.

Solution At a fixed instant t, position of the wave center in the x-axis is given by

$$\overline{x}(t) = \overline{\overline{v_0}} + t\overline{\overline{v}} \tag{9.54}$$

Now a single overbar denotes averaging in space (over the x), and a double overbar denotes averaging in spatial frequency (over the k); $v_0(k)$ is the group velocity in the initial spatial spectrum $U_0(k)$, and $v(k)$ is the group velocity in a medium. In pure dispersive media, spectral amplitudes are conserved, $A(t, k) = A_0(k) = |U_0(k)|$, and (9.54) shows again that the wave center moves with a constant velocity.

Whitham's equation takes the form

$$\frac{\partial k}{\partial t} + v(k)\frac{\partial k}{\partial x} = 0 \tag{9.55}$$

and characteristics are the straight lines given by

$$x = \xi + tv[k(\xi)] \tag{9.56}$$

Here, ξ is the point where the characteristic intersects the axis $t = 0$, and $k(\xi)$ is the initial spatial distribution of wave numbers at $t = 0$. Originally, these equations have been given in [2].

Problem 9.4

Find the exact wave function for a free particle.

Solution In a space-wave number domain, for Schrödinger's equation with $V(x) = 0$, dispersion relations (9.53) are as follows:

$$\omega = \frac{\hbar}{2m}k^2 \quad \text{and} \quad v(k) = \frac{d\omega}{dk} = \frac{\hbar}{m}k$$

Therefore, the second Fourier integral (9.52) takes the form

$$\psi(x, t) = \frac{1}{2\pi}\int_{-\infty}^{\infty} U_0(k) \exp i\left(\frac{\hbar t}{2m}k^2 - kx\right) dk$$

Here, $U_0(k)$ is the initial spatial spectrum at $t = 0$. For the Gaussian initial packet of a length L:

$$\psi_0(x) = \frac{1}{\sqrt{L}} e^{-ik_0x - x^2/2L^2} \tag{9.57}$$

we have

$$U_0(k) = \sqrt{2\pi}\, L e^{-(k-k_0)^2 L^2/2}$$

and the integral can be evaluated. Then, we obtain the wave function for a free particle [9]

$$\psi(x, t) = \frac{1}{\sqrt{L - i\frac{\hbar t}{mL}}} \exp\left[-\frac{\frac{x^2}{2L} + iLk_0\left(x - k_0\frac{\hbar}{2m}t\right)}{L - i\frac{\hbar t}{mL}} \right] \tag{9.58}$$

As in (9.37), the wave packet broadens when propagating, and for the instant t, its effective length is

$$L(t) = \sqrt{L^2 + \frac{\hbar^2 t^2}{m^2 L^2}}$$

Now the length $L(t)$ is considered at a fixed t instead of the duration $T(x)$ at a fixed x.

Problem 9.5

Show that Whitham's method in the form (9.55) and (9.56) gives the same exact wave function for a free particle.

Solution For the initial wave function $\psi_0(\xi)$ given in (9.57), the complex wave number is $k(\xi) = k_0 - i\xi/L^2$, and for $v(k) = k\hbar/m$, (9.56) takes the form

$$x = \xi + \frac{\hbar t}{m}\left(k_0 - i\frac{\xi}{L^2}\right)$$

Therefore, we find

$$\xi(x, t) = \frac{x - \frac{\hbar t}{m}k_0}{1 - i\frac{\hbar t}{mL^2}}$$

$$k(x, t) = k_0 - \frac{i}{L^2}\frac{x - \frac{\hbar t}{m}k_0}{1 - i\frac{\hbar t}{mL^2}}$$

$$\theta(x, t) = k_0 x - \frac{i}{2L}\frac{\left(x - \frac{\hbar t}{m}k_0\right)^2}{L - i\frac{\hbar t}{mL}} + g(t)$$

The complex phase $\theta(x, t)$ is the integral of $k(x, t)$ with respect to x, and the function $g(t)$ should be found for the normalized wave function $\psi(x, t) = e^{-i\theta(x,t)}$ that conserves energy at each t. Then we come to the Gaussian amplitude $|\psi(x, t)|$ of the same length $L(t)$ as in (9.58).

Problem 9.6

Whitham's assumption is that the dispersion relation (9.24) derived for the constant ω and k is also valid for the slowly varying local $\omega(x, t)$ and $k(x, t)$. Using the SPA and the spectral phase (9.14), reassert this assumption for uniform dispersive media. Generalize this approach for nonuniform media. Also, explain exact wave function in Problem 9.5.

Solution According to the SPA (3.10), for the phase (9.14), the AW is as follows:

$$w(x, t) = \frac{1}{2\pi}\int_{-\infty}^{\infty} A(x, \omega)e^{-i[\psi_0(\omega)+xk(\omega)-\omega t]}\, d\omega$$
$$\sim \sqrt{\frac{-i}{2\pi\left[\psi_0''(x, \omega_s) + xk''(\omega_s)\right]}}\, A(x, \omega_s)e^{-i[\psi_0(\omega_s)+xk(\omega_s)-\omega_s t]}$$

Here the stationary point ω_s depending on x and t is defined by

$$-\frac{d\phi}{d\omega_s} = \tau_0(\omega_s) + x\tau_k(\omega_s) - t = 0 \tag{9.59}$$

and the phase of the AW $w(x, t)$ is given by

$$\phi(x, t) = -[\psi_0(\omega_s) + xk(\omega_s) - \omega_s t]$$

Therefore, differentiating $\phi(x, t)$ with respect to t and x, we find

$$\begin{aligned}\frac{\partial\phi}{\partial t} &= \omega_s - [\tau_0(\omega_s) + x\tau_k(\omega_s) - t]\frac{\partial\omega_s}{\partial t} = \omega_s \\ -\frac{\partial\phi}{\partial x} &= k(\omega_s) + [\tau_0(\omega_s) + x\tau_k(\omega_s) - t]\frac{\partial\omega_s}{\partial x} = k(\omega_s)\end{aligned} \tag{9.60}$$

where the bracket is zero because of (9.59) for the stationary point ω_s. As in (3.11), the local instantaneous frequency $\partial\phi/\partial t$ is the stationary point ω_s. Also, the local wave number $-\partial\phi/\partial x$ is $k(\omega_s)$. Since (9.24) is valid for any frequency including ω_s, this reasserts that the dispersion relation is true for the local frequency and wave number, both depending on x and t. So, Whitham's method directly results from the SPA.

For nonuniform media, using (9.20) with $\chi = k(x, \omega)$, we come to the same relations where $xk(\omega)$ is replaced by $\int_0^x k(x, \omega)\, dx$. Therefore, equations (9.60) remain valid. It may also be noted that the complex phase (9.25) relates to the saddle point method generalizing the SPA.

The SPA is exact for signals with quadratic phases. For a free particle in a space-wave number domain, dispersion relation is quadratic $\omega = \frac{\hbar}{2m} k^2$, and the phase is quadratic also. Therefore, Whitham's method gives exact wave function.

Problem 9.7

Show that (9.46) is equivalent to (9.45) and both are of the first order with respect to ε.

Solution Differentiating (9.46) with respect to x and t and neglecting the derivatives of a slow amplitude $a(x, t)$, we have

$$\frac{\partial^2 w}{\partial x^2} + \frac{1}{c(a)} \frac{\partial^2 w}{\partial t\, \partial x} = 0 \qquad \frac{\partial^2 w}{\partial t\, \partial x} + \frac{1}{c(a)} \frac{\partial^2 w}{\partial t^2} = 0$$

Therefore, eliminating the mixed derivative, we come to (9.45) for nondispersive media:

$$\frac{\partial^2 w}{\partial x^2} - \frac{1}{c^2(a)} \frac{\partial^2 w}{\partial t^2} = 0$$

So, (9.45) and (9.46) are equivalent for nondispersive media and for slowly varying amplitudes. For the velocity $c(a)$ given in (9.45), (9.46) is of the first order with respect to ε. However, (9.45) is also of the first order because we have neglected the amplitude derivative.

Problem 9.8

Deduce duration (9.37) from the characteristic (9.36).

Solution For $\omega(\xi) = \omega_0 + i\xi/T_0^2$ and for a small ξ, (9.36) becomes linear and gives

$$\xi = \frac{t - (x + x_0)\sqrt{m/2(\hbar\omega_0 - V_0)}}{1 - i(x + x_0)F/T_0^2}$$

where the constant F is given by

$$F = \frac{\hbar\sqrt{m/2}}{2(\hbar\omega_0 - V_0)^{3/2}}$$

Therefore, as in Section 9.3.2, we find

$$\omega(x, t) = \omega_0 + i\frac{t - (x + x_0)\sqrt{m/2(\hbar\omega_0 - V_0)}}{T_0^2 - i(x + x_0)F}$$

$$\theta(x, t) = \omega_0 t + i\frac{(t - (x + x_0)\sqrt{m/2(\hbar\omega_0 - V_0)})^2}{2\left(T_0^2 - i(x + x_0)F\right)}$$

Disregarding normalization, the amplitude $a(x, t)$ relates to the complex phase as $\ln a(x, t) = \mathrm{Re}[i\theta(x, t)]$. Therefore, eliminating the imaginary part in the denominator of θ, we find

$$\ln a(x, t) = -\frac{(t - (x + x_0)\sqrt{m/2(\hbar\omega_0 - V_0)})^2}{2T^2(x)}$$

where duration $T(x)$ is given by (9.37).

References

[1] Vakman, D., "Analytic Waves," *Intern. Jour. Theor. Phys.*, Vol. 36, 1997, pp. 227–247.

[2] Whitham, G. B., *Linear and Nonlinear Waves*, New York: John Wiley, 1974.

[3] Kotler, Z., and A. Nitzan, "Traversal Time for Tunelling: Local Aspects," *Jour. Chem. Phys.*, Vol. 88, 1988, pp. 3871–3878.

[4] Landauer, R., and T. Martin,"Barrier Interaction Time in Tunelling," *Review of Modern Physics*, Vol. 68, 1994, pp. 217–228.

[5] Drazin, P. G., and R. S. Lohnson, *Solitons: An Introduction*, Cambridge: Cambridge University Press, 1992.

[6] Hahn, S. L., "Multidimensional Complex Signals with Single–Orphant Spectra," *Proc. IEEE*, Vol. 80, 1992, pp. 1287–1300.

[7] Heading, J., *An Introduction to Phase-Integral Methods*, London: Methuen, New York: John Wiley, 1962.

[8] Kay, I., and R. A. Silverman,"On the Uncertainty Relation for Real Signals," *Information and Control*, Vol. 1, 1957, pp. 64–75.

[9] Saxon, D. S., *Elementary Quantum Mechanics*, New York: McGraw-Hill, 1968.

About the Author

David Vakman was born in Moscow, Russia. He received B.S., M.S., and Ph.D. degrees in radio and electrical engineering from Moscow Aviation University in 1949, 1961, and 1970, respectively.

Since 1949, he has worked as a research scientist in both military and industrial electronics. His current research interests include oscillation and wave theory and signal processing. He is the author or co-author of a number of books in Russian and in English. He has also published approximately 100 articles in scientific journals. His e-mail address is DVakman@aol.com.

Index

The Artech House Signal Processing Library

Alexander Poularikas, Series Editor

Hilbert Transforms in Signal Processing, Stefan L. Hahn

Phase and Phase-Difference Modulation in Digital Communications, Yuri Okunev

Principles of Signals and Systems: Deterministic Signals, Bernard Picinbono

Signals. Oscillations, and Waves: A Modern Approach, David Vakman

Statistical Signal Characterization, Herbert L. Hirsch

Statistical Signal Characterization Algorithms and Analysis Programs, Herbert L. Hirsch

Surface Acoustic Waves for Signal Processing, Michel Feldman and Jeannine Henaff